NATIONAL ACADEMIES *Sciences Engineering Medicine*

NATIONAL ACADEMIES PRESS
Washington, DC

Greenhouse Gas Emissions from Wildland Fires

Toward Improved Monitoring, Modeling, and Management

Rachel Silvern, *Rapporteur*

Committee on Greenhouse Gas Emissions from Wildland Fires: Toward Improved Monitoring, Modeling, and Management

Board on Atmospheric Sciences and Climate

Board on Agriculture and Natural Resources

Polar Research Board

Division on Earth and Life Studies

Proceedings of a Workshop

NATIONAL ACADEMIES PRESS 500 Fifth Street, NW Washington, DC 20001

This activity was supported by a contract between the National Academy of Sciences and SATManagement, a wholly owned subsidiary of the Environmental Defense Fund, and the National Academy of Sciences Arthur L. Day Fund. Any opinions, findings, conclusions, or recommendations expressed in this publication do not necessarily reflect the views of any organization or agency that provided support for the project.

International Standard Book Number-13: 978-0-309-71553-9
International Standard Book Number-10: 0-309-71553-9
Digital Object Identifier: https://doi.org/10.17226/27473

This publication is available from the National Academies Press, 500 Fifth Street, NW, Keck 360, Washington, DC 20001; (800) 624-6242 or (202) 334-3313; http://www.nap.edu.

Copyright 2024 by the National Academy of Sciences. National Academies of Sciences, Engineering, and Medicine and National Academies Press and the graphical logos for each are all trademarks of the National Academy of Sciences. All rights reserved.

Printed in the United States of America.

Suggested citation: National Academies of Sciences, Engineering, and Medicine. 2024. *Greenhouse Gas Emissions from Wildland Fires: Toward Improved Monitoring, Modeling, and Management: Proceedings of a Workshop.* Washington, DC: The National Academies Press. https://doi.org/10.17226/27473.

The **National Academy of Sciences** was established in 1863 by an Act of Congress, signed by President Lincoln, as a private, nongovernmental institution to advise the nation on issues related to science and technology. Members are elected by their peers for outstanding contributions to research. Dr. Marcia McNutt is president.

The **National Academy of Engineering** was established in 1964 under the charter of the National Academy of Sciences to bring the practices of engineering to advising the nation. Members are elected by their peers for extraordinary contributions to engineering. Dr. John L. Anderson is president.

The **National Academy of Medicine** (formerly the Institute of Medicine) was established in 1970 under the charter of the National Academy of Sciences to advise the nation on medical and health issues. Members are elected by their peers for distinguished contributions to medicine and health. Dr. Victor J. Dzau is president.

The three Academies work together as the **National Academies of Sciences, Engineering, and Medicine** to provide independent, objective analysis and advice to the nation and conduct other activities to solve complex problems and inform public policy decisions. The National Academies also encourage education and research, recognize outstanding contributions to knowledge, and increase public understanding in matters of science, engineering, and medicine.

Learn more about the National Academies of Sciences, Engineering, and Medicine at **www.nationalacademies.org**.

Consensus Study Reports published by the National Academies of Sciences, Engineering, and Medicine document the evidence-based consensus on the study's statement of task by an authoring committee of experts. Reports typically include findings, conclusions, and recommendations based on information gathered by the committee and the committee's deliberations. Each report has been subjected to a rigorous and independent peer-review process and it represents the position of the National Academies on the statement of task.

Proceedings published by the National Academies of Sciences, Engineering, and Medicine chronicle the presentations and discussions at a workshop, symposium, or other event convened by the National Academies. The statements and opinions contained in proceedings are those of the participants and are not endorsed by other participants, the planning committee, or the National Academies.

Rapid Expert Consultations published by the National Academies of Sciences, Engineering, and Medicine are authored by subject-matter experts on narrowly focused topics that can be supported by a body of evidence. The discussions contained in rapid expert consultations are considered those of the authors and do not contain policy recommendations. Rapid expert consultations are reviewed by the institution before release.

For information about other products and activities of the National Academies, please visit www.nationalacademies.org/about/whatwedo.

COMMITTEE ON GREENHOUSE GAS EMISSIONS FROM WILDLAND FIRES: TOWARD IMPROVED MONITORING, MODELING, AND MANAGEMENT

LORETTA J. MICKLEY (*Chair*), Harvard University
SALLY ARCHIBALD, University of the Witwatersrand
CHRIS (FERN) FERNER, Esri (*Formerly*)
NANCY FRENCH, Michigan Technological University
DON HANKINS, California State University, Chico
WERNER KURZ, Canadian Forest Service, Natural Resources Canada (*Retired*)
JAMES RANDERSON (NAS), University of California, Irvine
BRENDAN ROGERS, Woodwell Climate Research Center
AMBER SOJA, National Aeronautics and Space Administration

Staff

RACHEL SILVERN, Program Officer, Board on Atmospheric Sciences and Climate (BASC)
KATRINA HUI, Associate Program Officer, BASC
KARA LANEY, Senior Program Officer, Board on Agriculture and Natural Resources
ANNIE MANVILLE, Program Assistant, BASC
RACHEL SANCHEZ, Senior Program Assistant, BASC (*until January 3, 2024*)
HUGH WALPOLE, Associate Program Officer, BASC (*until July 2023*)

BOARD ON ATMOSPHERIC SCIENCES AND CLIMATE

MARY GLACKIN (*Chair*), The Weather Company, an IBM Business (*Retired*)
JOSEPH ÁRVAI, University of Southern California
CYNDI ATHERTON, Heising-Simons Foundation (*Formerly*)
ELIZABETH BARNES, Colorado State University
BRAD R. COLMAN, American Meteorological Society
BART CROES, California Air Resources Board (*Retired*)
MINGHUI DIAO, San Jose State University
NEIL DONAHUE, Carnegie Mellon University
LESLEY-ANN DUPIGNY-GIROUX, University of Vermont
EFI FOUFOULA-GEORGIOU (NAE), University of California, Irvine
KEVIN GURNEY, Northern Arizona University
MARIA CARMEN LEMOS (NAS), University of Michigan
ANDREA LOPEZ LANG, University of Wisconsin–Madison
ZHANQING LI, University of Maryland
AMY McGOVERN, University of Oklahoma
LINDA MEARNS, National Center for Atmospheric Research
JONATHAN PATZ (NAM), University of Wisconsin–Madison
KEVIN REED, Stony Brook University
JAMES MARSHALL SHEPHERD (NAS/NAE), University of Georgia
ARADHNA TRIPATI, University of California, Los Angeles
BERNEDETTE WOODS PLACKY, Climate Central

Staff

ELIZABETH EIDE, Acting Director
MAGGIE WALSER, Acting Director
KATELYN CREWS, Program Assistant
APURVA DAVE, Senior Program Officer
MORGAN DISBROW-MONZ, Program Officer
KATRINA HUI, Associate Program Officer
ANNIE MANVILLE, Program Assistant
BRIDGET McGOVERN, Program Officer
APRIL MELVIN, Senior Program Officer
LINDSAY MOLLER, Senior Program Assistant
RACHEL SILVERN, Program Officer
STEVEN STICHTER, Senior Program Officer

BOARD ON AGRICULTURE AND NATURAL RESOURCES

JILL J. McCLUSKEY (*Chair*), Washington State University
AMY W. ANDO, The Ohio State University
ARISTOS ARISTIDOU (NAE), Biomason
BRUNO BASSO, Michigan State University
BERNADETTE DUNHAM, George Washington University
JESSICA E. HALOFSKY, U.S. Forest Service
ERMIAS KEBREAB, University of California, Davis
MARTY D. MATLOCK, University of Arkansas
JOHN P. MCNAMARA, Washington State University
NAIMA MOUSTAID-MOUSSA, Texas Tech University
V. ALARIC SAMPLE, George Mason University
ROGER E. WYSE, Spruce Capital Partners

Staff

ROBIN SCHOEN, Director
CAMILLA YANDOC ABLES, Senior Program Officer
MALIA BROWN, Program Assistant
KARA LANEY, Senior Program Officer
ALBARAA SARSOUR, Program Officer
SAMANTHA SISANACHANDENG, Senior Program Assistant

Reviewers

This Proceedings of a Workshop was reviewed in draft form by individuals chosen for their diverse perspectives and technical expertise. The purpose of this independent review is to provide candid and critical comments that will assist the National Academies of Sciences, Engineering, and Medicine in making each published proceedings as sound as possible and to ensure that it meets the institutional standards for quality, objectivity, evidence, and responsiveness to the charge. The review comments and draft manuscript remain confidential to protect the integrity of the process.

We thank the following individuals for their review of this proceedings:

LORI DANIELS, University of British Columbia
PETER FRUMHOFF, Woodwell Climate Research Center and Harvard University Center for the Environment
DON HANKINS, California State University, Chico
DAN JAFFE, University of Washington
LORETTA MICKLEY, Harvard University

Although the reviewers listed above provided many constructive comments and suggestions, they were not asked to endorse the content of the proceedings nor did they see the final draft before its release. The review of this proceedings was overseen by **DAVID ALLEN (NAE)**, The University of Texas at Austin. He was responsible for making certain that an independent examination of this proceedings was carried out in accordance with standards of the National Academies and that all review comments were carefully considered. Responsibility for the final content rests entirely with the rapporteur and the National Academies.

Contents

OVERVIEW 1

INTRODUCTION 3
Framing the Workshop, 4

**BIOMES VULNERABLE TO WILDLAND FIRES AND IMPLICATIONS
FOR GREENHOUSE GAS EMISSIONS**..9
Roles of Fire in Presently Vulnerable Biomes and the Associated Net Greenhouse
   Gas Emissions, 9
Management of Fires and Ecosystems and Implications for Greenhouse Gas
   Emissions: Recent Past and Current, 17

**OBSERVING AND MODELING WILDLAND FIRES AND THEIR
GREENHOUSE GAS EMISSIONS: OPPORTUNITIES AND CHALLENGES**.....26
Observation-Based Approaches for Quantifying Emissions from Wildland Fires, 26
Modeling Emissions from Wildland Fires, 32

FUTURE MANAGEMENT TO SUPPORT NET-ZERO TARGETS 40
Accounting for Wildfire Emissions in National Reporting and Net-Zero Targets, 40
Opportunities to Reduce Future Wildfire Emissions in Different Biomes, 44
The Solution Space and Examples of Next Steps: Forest Management of Tomorrow
   and Livable Emissions, 54

CLOSING THOUGHTS 58

REFERENCES..61

APPENDIXES

A: STATEMENT OF TASK 71

B: BIOGRAPHICAL SKETCHES OF COMMITTEE MEMBERS........................72

C: WORKSHOP AGENDA 76

Overview

Fire is a natural part of global ecosystems that has shaped the landscapes humans live on today. However, human influences from land management and development practices and climate change have resulted in wildland fires that burn hotter, more frequently, over larger areas, or in some locations where fires had not typically been a part of the ecosystem. Fires transfer carbon between terrestrial pools and between the land and atmosphere through emissions of greenhouse gases (GHGs) and other gases and particles. Direct and indirect (i.e., climate change) human-driven changes in wildland fire regimes have the potential to increase GHG emissions at a scale that could inhibit global efforts to achieve "net-zero" GHG emissions in the coming decades. To inform land management and policy decisions, it is important to better understand future changes in wildland fires and their GHG emissions as well as effective management strategies that could limit potential GHG emissions and preserve ecosystems.

The National Academies of Sciences, Engineering, and Medicine convened a workshop on September 13–15, 2023, to identify opportunities to improve measurements and model projections of GHG emissions from wildland fires and discuss management practices that could be incorporated into current and future action plans. Workshop discussions sought to identify how tools and changes in land stewardship could inform and enhance strategies to limit wildland fire GHG emissions and associated threats to achieving net-zero emission targets.

Climate change is fundamentally changing ecosystems and their potential future fire behavior and effects. Many participants emphasized the importance of prioritizing reductions in anthropogenic GHG emissions to reduce wildland fire emissions driven by climate change. Costs of fire management strategies to reduce GHG emissions may compete with other government-funded priorities, and better tools may be needed to help decision makers consider effective investments.

A clear theme throughout the workshop was the importance of learning from historic and current practices of Indigenous peoples, reintroducing cultural burning and cultural land management practices, and including and elevating the leadership of Indigenous peoples in all stages of fire management. Engaging local communities who live in and manage ecosystems vulnerable to wildfire and benefit from land use is an important part of identifying and implementing appropriate intervention strategies.

Discussions were organized around three global biomes—temperate, Arctic/boreal, and tropical—where historical fire regimes and the carbon balance have been disrupted due to human-driven climate and land use changes. Many participants emphasized the heterogeneity in global ecosystems, fire regimes, and the associated management practices that are appropriate. For example, while there are potential opportunities to increase fire suppression in some ecosystems such as the northern boreal, in many other biomes, a legacy of fire suppression may actually be contributing to the extreme fire conditions resulting in megafires (e.g., western United States).

Emission inventories are important tools for estimating GHG emissions from wildfires and ecosystem removals (i.e., uptake and sequestration) of carbon. Across ecosystems, there are limited data on all aspects of fire emissions, with important implications for GHG emissions. In particular, the impact of fires on soils, peat, and permafrost could be significant, given that such fires release to the atmosphere carbon that has been stored for decades to centuries. Participants also discussed tools and models that could help managers identify where fire mitigation strategies are climate-effective.

Discussions centered around the importance of regionally differentiated, ecosystem-appropriate mitigation strategies ranging from prescribed fires, fuel management, water table management, reduction in human ignitions, and targeted suppression, among others. Several participants recognized the co-benefits and trade-offs of different management approaches at local scales—for example, thinning in dry ecosystems can increase water availability and ecosystem health. There are opportunities to consider fuel reduction strategies that instead of burning, utilize biomass in ways that can benefit the bioeconomy, reduce fuel management costs, and reduce smoke and health impacts.

Moving forward, the challenge for decision makers and managers will be scaling up mitigation actions to have meaningful impacts while also considering trade-offs at large scales. As an example, implementing management strategies discussed at the workshop may mean shifting public acceptance of limited smoke from prescribed burning as an alternative to uncontrolled wildfire smoke. At the same time, local community concerns—for example, concerns about escaped prescribed fires in the wildland–urban interface—are important to address. Broadly, workshop discussions highlighted the wide range of available management solutions that would reduce GHG emissions and increase the resilience of vulnerable ecosystems.

Introduction

The human, environmental, and economic damages from wildland fires are a growing global challenge as wildland fires burn hotter, more frequently, and over larger areas and occur in systems where fire had not historically been part of the ecosystem. Today, climate change and certain human land management and development practices are exacerbating the conditions and locations where wildland fires develop, and the resulting intensity of those events, which can overwhelm the resources available for management (UNEP, 2022). While landscape fires are a natural part of healthy, evolving ecosystems, large, uncontrolled wildland fires can have devastating consequences on human health, communities, and biodiversity. Ecosystems store large amounts of carbon as above- and belowground biomass, and wildland fires in peatlands, tundra, forests, grasslands, and other systems can emit large amounts of carbon dioxide (CO_2) and other greenhouse gases (GHGs) to the atmosphere.

Fires have been a natural part of the Earth system for hundreds of millions of years and have played a critical role in shaping global ecosystems (Bond et al., 2005). Climate affects fire regimes through its control on weather and interactions with vegetation productivity and structure—for example, through drought or fuel desiccation or flammability—at the global, regional, and local scales (Jia et al., 2019). Today, humans are the main source of fire ignitions globally (Bowman et al., 2017; Harris et al., 2016), and humans also influence fires by extinguishing them, reducing their spread, and managing their fuel sources, as well as broadly changing land use.

Fire transfers carbon between the land and atmosphere through emissions of GHGs, carbon monoxide (CO), particulate matter, and other gases; fire also transfers carbon between different terrestrial pools from live to dead biomass, pyrogenic carbon, and soil. After a fire burns, vegetation can recover and regrow to sequester carbon from the atmosphere as biomass over years to decades, depending on the biome type (Landry and Matthews, 2016). However, human-driven changes to climate and land use have made biomes vulnerable to changes in fire regimes and are disrupting the carbon balance of these systems.

Future changes in climate—for example, increasing temperatures, decreasing precipitation, and changing regional weather extremes—are expected to increase the risk and severity of wildfires in many global biomes (Jia et al., 2019). Continued climate-driven changes in wildland fire regimes have the potential to increase GHG emissions at a scale that could counteract global reductions in anthropogenic GHG emissions that have been made to achieve the 2°C temperature target set forth by the Paris Agreement.[1] There are large uncertainties in measurements and models of GHG emissions from wildland fires

---

[1] The Paris Agreement is an international treaty on climate change, adopted by 196 Parties at the United Nations Climate Change Conference (COP21) in 2015. Its overarching goal is to hold "the increase in the global average temperature to well below 2°C above pre-industrial levels" and pursue efforts "to limit the temperature increase to 1.5°C above pre-industrial levels."

today, and projections of how wildland fires and their emissions will change on decadal to century timescales. Better understanding of how the feedbacks between wildland fires, their GHG emissions, and climate change could effect global efforts to achieve net-zero GHG emissions in the coming decades would be useful. At the same time, it is important to develop and implement management strategies, including from Indigenous knowledge and practices, that could limit potential GHG emissions from wildland fires while also addressing the immediate needs of affected communities and ecosystems.

In September 2023, the Board on Atmospheric Sciences and Climate together with the Board on Agriculture and Natural Resources and the Polar Research Board of the National Academies of Sciences, Engineering, and Medicine convened a workshop on GHG emissions from wildland fires. The workshop identified opportunities to improve measurements and model projections of GHG emissions from wildland fires, considered how changes in emissions from these fires could affect our ability to achieve "net-zero" GHG emission targets, and discussed management practices that could be incorporated in current and future action plans. See Appendix A for the statement of task to the workshop planning committee; Appendix B for biographical sketches of the planning committee members; and Appendix C for the workshop agenda. Box 1 provides key terminology used throughout the proceedings.

The structure of the proceedings largely follows the workshop agenda (see Appendix C). The 3 days of the workshop were organized around three broad themes:

1. Biomes vulnerable to wildland fires and implications for GHG emissions,
2. Opportunities and challenges for observing and modeling wildland fires and their GHG emissions, and
3. Future management to support net-zero targets.

This proceedings summarizes workshop presentations and discussions in the plenary and breakout sessions. This proceedings has been prepared by the workshop rapporteur as a factual summary of what occurred at the workshop. The views contained in the proceedings are those of individual workshop participants and do not necessarily represent the views of all workshop participants, the planning committee, or the National Academies of Sciences, Engineering, and Medicine. Funding for this workshop was provided by SATManagement, a wholly owned subsidiary of the Environmental Defense Fund, and the National Academy of Sciences Arthur L. Day Fund.

FRAMING THE WORKSHOP

To begin the workshop, **Scott Stephens, University of California, Berkeley**, centered the relationship between Indigenous peoples and fire. Stephens shared a perspective on fire from Val Lopez, chairman of the Amah Mutsun Tribal Band: "Fire is a gift from creator for the stewardship of the land." Stephens challenged the perception of fire as something only to fear as a destructive force and suggested that viewing fire as a gift is fundamental to any forward progress.

> **BOX 1**
> **Key Concepts and Terminology**
>
> **Environmental stewardship**: The responsible use and protection of the natural environment through conservation and sustainable practices to enhance ecosystem resilience and human well-being (Chapin et al., 2010).
>
> **Fuel**: Live and dead vegetation biomass that burns, including aboveground live, dead surface material, and ground-layer organic matter.
>
> **Fuels or fire management**: Practices and strategies, including land management fires and cultural burning, used as a tool for achieving resilient ecosystems.
>
> **Greenhouse gases (GHGs)**: Gases that trap heat in the atmosphere. Water vapor, carbon dioxide (CO_2), methane, nitrous oxide, and ozone are the primary GHGs in Earth's atmosphere (IPCC, 2021). Carbon-containing GHGs primarily responsible for warming (e.g., CO_2) were the focus of this workshop.
>
> **Indigenous (fire) stewardship**: Indigenous fire stewardship is the concept used by various Indigenous, Aboriginal, and tribal peoples to explain the intergenerational teachings of fire-related knowledge, beliefs, and practices among fire-dependent cultures regarding fire regimes, fire effects, and the role of cultural burning in fire-prone ecosystems and habitats (Maclean et al., 2023).
>
> **Land use legacy**: Land condition is a result of past land use and driven by both natural state factors and human land stewardship practices.
>
> **Net zero**: Requires cutting GHG emissions to as close to zero as possible, with any remaining emissions re-absorbed from the atmosphere.[a]
>
> **Wildland**: Area in which contemporary human development is essentially nonexistent except for roads, railroads, power lines, and similar transportation or utility structures.[b]
>
> **Wildland fires**: Fires that originate in the "wildlands," as opposed to structure fires and fires occurring in built environments. Includes planned and unplanned burns, cultural burning, management fire, wildfire, rangeland burning, as well as escaped agricultural and other planned burns.
>
> [a] See https://www.un.org/en/climatechange/net-zero-coalition.
> [b] See https://lod.nal.usda.gov/nalt/302858.

Steve Hamburg and **Ann Bartuska, Environmental Defense Fund (EDF),** shared their motivation for sponsoring the workshop. There is a knowledge gap in current understanding of the extent of wildland fires, their GHG emissions, and the implications for reaching net-zero GHG emissions. Hamburg shared EDF's interest in creating a scientific foundation to make decisions about opportunities to reduce GHG emissions from wildfires, particularly wildfires in unmanaged lands. Hamburg charged the participants to think boldly about the wildfire challenge and to explore what is known, where there are knowledge gaps, and a possible path forward built upon a solid foundation of science. Bartuska recognized the importance of highlighting and learning from Indigenous and cultural burning practices as part of the workshop and discussion of the solution space, and noted that the timing of the workshop was shortly before the release of the Wildland Fire Mitigation and Management Commission report (2023), which outlined recommendations to Congress to address the U.S. wildfire crisis.

Workshop Scope

On behalf of the workshop planning committee, **Nancy French, Michigan Technological University,** provided framing and motivation for workshop discussions. French emphasized the key charge to workshop participants: Identify examples of gaps and opportunities in data and models, and changes in land stewardship, that can inform and enhance strategies to limit wildland fire GHG emissions and associated threats to net-zero emission targets.

Biomass burning is an integral force in the Earth system at the global scale, and fire is a management tool used at local and landscape scales. While smoke from fire can affect air quality and human health, the workshop focused on fire as a source of carbon and other GHG emissions, though air quality and GHG emissions are intertwined. The workshop focused on managed burning (where fires are allowed to burn naturally and only suppressed under defined management conditions), unintentional fires that are then managed, and unmanaged fires, because these fire types have the largest potential impacts on GHG emissions. Biomass burning from sources such as wood gathering, cookstoves, and cropland burning were not the focus and are less significant from the perspective of GHG emissions.

The workshop considered fire as a key factor in the carbon cycle at all scales from global to local and short term to long term. Fire has impacts on the carbon cycle of terrestrial ecosystems at multiple timescales: first, there are direct emissions of smoke; then, there may be delayed emissions as the system adjusts to new conditions; and finally, there is subsequent regrowth and ecosystem changes, which can transform carbon cycling. French explained that ecosystems where recovery post-fire is long—tens to hundreds of years—should be the focus for considering the impact of fire on the global carbon budget.

The workshop centered on ecosystems vulnerable to changes in fire conditions and regimes due to present and future direct and indirect (i.e., climate change-driven) human influences. In particular, sessions focused on archetypes of fires in three broad biomes, which each have different main drivers of change (Figures 1 and 2):

1. **Temperate biomes**: Change is due to a combination of climate, historical land use legacies, and current fire and ecosystem management practices;

2. **Arctic/boreal biomes**: Change is dominated by climate (indirect human influence) with direct human influence presently limited, but emerging; and
   3. **Tropical biomes**: Change is dominated by economically driven systems and fire management with limited climate influence.

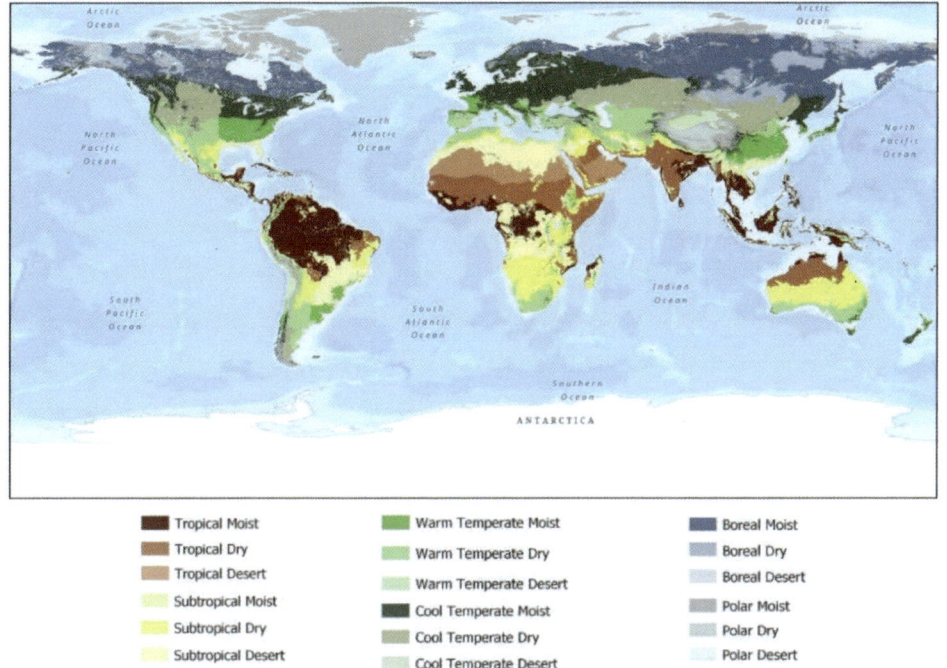

FIGURE 1 World Climate Regions shown as geospatial integration of World Temperature Domains and World Moisture Domains. Topical regions are shown in brown, subtropical regions shown in yellow, temperate regions shown in green, boreal regions shown in blue, and polar regions shown in grey. SOURCE: Sayre et al. (2020).

Workshop discussions focused on landscapes where there has been a disruption in fire regime in the past and present, including where fire is common for ecosystem maintenance or landscapes, where historical fire regimes and cultural practices have been disrupted due to colonialism, or other land use forces (Figure 2). Ecosystems where the carbon cycle is being drastically impacted—for example, landscapes where fire is novel, the fire regime is changing, or old carbon stocks are being disrupted—were of particular interest, rather than grassland and savannah landscapes that experience rapid revegetation following fire. French identified the challenge of addressing fires across this broad diversity of biomes because they require regionally differentiated and ecologically appropriate solutions that account for the biome-specific fire regime, plant and ecosystem adaptations and vulnerabilities, and the legacy of people on the land.

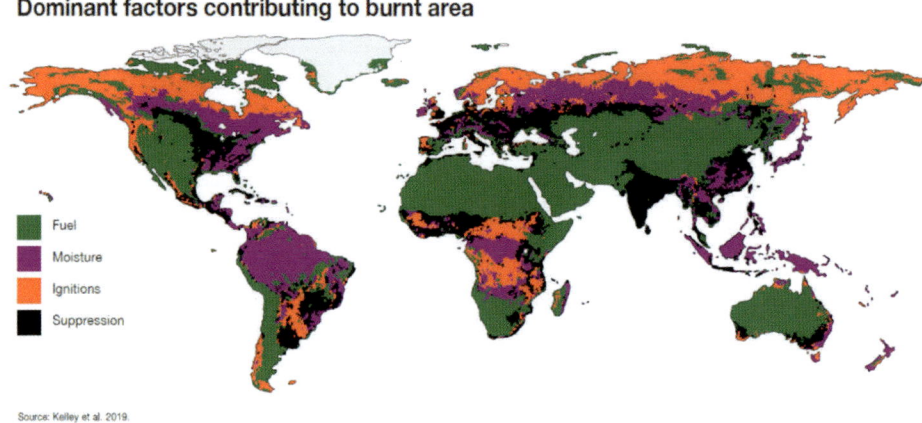

FIGURE 2 Dominant factors contributing to burnt area during the height of the fire season (the month of the year that, on average, experiences the highest burnt area). Regions in green are predominantly sensitive to the presence of available fuel to carry fire; orange regions are predominantly sensitive to sources of successful ignition; purple regions are predominantly sensitive to fuel moisture, meaning that while fuel and ignitions may be sufficient, fuels are often not dry enough to combust; and black regions are predominantly sensitive to curtailment of fires through fire prevention, active suppression, or land-use fragmentation. SOURCE: GRID-Arendal, https://www.grida.no/resources/15556.

Biomes Vulnerable to Wildland Fires and Implications for Greenhouse Gas Emissions

The first day of the workshop considered how human activity—through anthropogenic climate change and land use change—have affected fire activity and greenhouse gas (GHG) emissions across different global ecosystems. Participants assessed how shifts in baseline fire regimes (e.g., through fire exclusion or inclusion) have perturbed fire activity and the carbon cycle (e.g., introducing fire to ecosystems with old, stored carbon) and the implications of increased frequency and severity of extreme wildfires for GHG emissions. Understanding current conditions of these ecosystems informed workshop discussions about feasible and effective solutions for management. Sessions first introduced conditions in currently vulnerable biomes, and second, considered current opportunities for management.

ROLES OF FIRE IN PRESENTLY VULNERABLE BIOMES AND THE ASSOCIATED NET GREENHOUSE GAS EMISSIONS

Across multiple sessions on the first day of the workshop, speakers introduced the three biomes of focus—temperate, Arctic/boreal, and tropical—and the factors making these ecosystems vulnerable to wildfire and significant emissions of GHGs today and in the future under climate change. Speakers considered the changes in net GHG emissions from fires and the necessary conditions to maintain these ecosystems in a resilient state.

Temperate Biomes

Several speakers provided historical context for the changing role of fire in temperate biomes—particularly in the western United States—and the ways in which the colonial policies of fire suppression have made these systems vulnerable. Using California as an example, **Scott Stephens, University of California, Berkeley,** noted that pre-1850 (statehood), approximately 1.8 million hectares (ha) burned annually, with about half the area burned coming from lightning and half from Indigenous burning (Stephens et al., 2007). **Hugh Safford, University of California, Davis** and **Vibrant Planet,** explained that the current burned area in California does not come close to an average year of area burned pre-1850 (Stephens et al., 2007). The policy of fire suppression over the last century is a major driver of ecological degradation in the state's ecosystems (Figure 3). **Karin Riley, U.S. Forest Service,** shared the example of low-elevation ponderosa pine forests in Montana that would have seen fire every 7–25 years prior to colonization. When Indigenous burning was removed as a process on the land, the European forestry model of suppression, which saw fire as a threat to timber resources, became the 20th-century practice of the U.S. Forest Service.

FIGURE 3 Example of significant landscape changes in western U.S. dry temperate forests between 1934 (top) and 2010 (bottom). This frequent fire system historically burned at an interval of 2–15 years, but after fire suppression policies over less than 100 years, the dry south slopes and ridges filled in with trees where flames can climb the layered subcanopy. Frequent low- or moderate-severity fires provided important local stabilizing feedback, improving the likelihood that subsequent fires would also be low or moderate severity. SOURCE: National Archives at Seattle (top); John F. Marshall (bottom).

Matthew Hurteau, University of New Mexico, emphasized that restoring the process of ecologically appropriate fire to temperate ecosystems is the key to maintaining stability for both carbon and forest cover. As a forest develops, it accumulates carbon over time until a process such as a high-severity wildfire decreases that live tree carbon. Hurteau introduced the carbon carrying capacity framework (i.e., a system's stable carbon stock over time) (Figure 4) in which, as long as prevailing climatic and natural process conditions remain consistent, the same amount of carbon will be sequestered over the same period of time. However, shifts in the prevailing climate or natural process conditions can result in temperate systems that store a lesser amount of carbon.

There are broadly two types of fire regimes in temperate forests, both of which are directly and indirectly influenced by humans: (1) climate driven and (2) human and climate driven (Figure 4). In the climate-driven regime, in general, vegetation is unavailable to burn most years because temperatures are too low or moisture is too high to support combustion, so only during hot, dry years is the system available to burn. In the human- and climate-driven regime, in general, temperate forests are available to burn every year during a fire season, and climate's effect on vegetation productivity and the amount and connectivity of the fuel present influence how often the system burns. Humans have influenced these regimes for millennia through purposeful ignitions to achieve management objectives—commonly referred to today as cultural burning—illustrating the impacts that fire stewardship has had across many ecosystems historically.

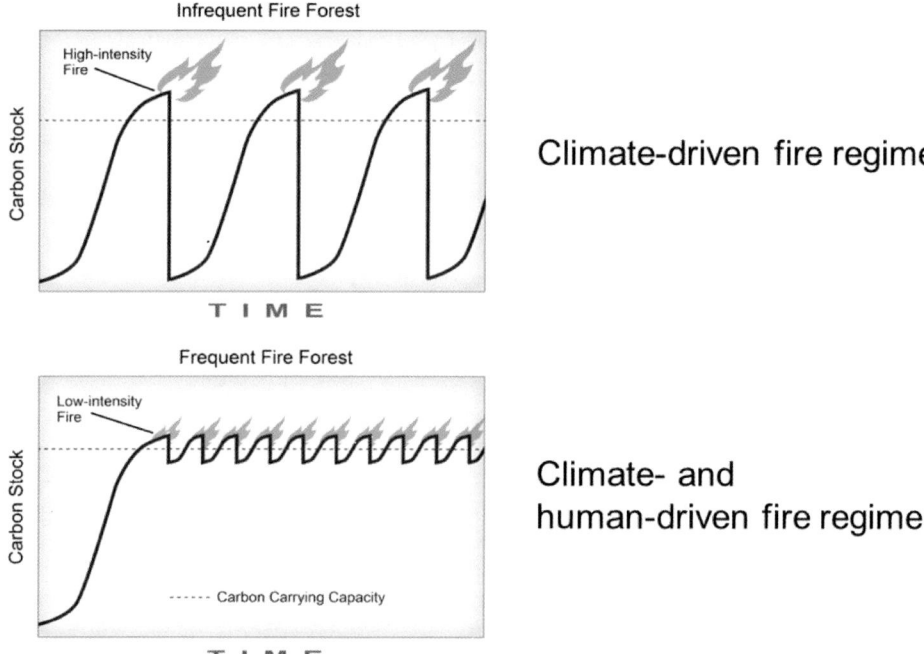

FIGURE 4 Framework of carbon carrying capacity for temperate ecosystems. As a forest develops, it accumulates carbon over time until the live tree carbon decreases due to a natural process (e.g., high-severity wildfire). In this carbon carrying capacity framework, after the disturbance event, as long as prevailing climatic and natural process conditions remain consistent, the ecosystem will sequester the same amount of carbon over the same time period (dashed horizontal lines). (Top) Climate-driven fire regime where in most years, vegetation is unavailable to burn due to low temperatures or moisture, so the system only burns during hot, dry years. (Bottom) Human- and climate-driven fire regime where there is an annual fire season where the effect of climate on vegetation productivity, the amount of fuel present, and its connectivity on the landscape determine how often the system can burn. If the prevailing climate or natural process conditions shift, the ecosystem can change to a lower carbon state (e.g., shrubland or a grassland). SOURCE: Hurteau (2013). Reproduced with permission of The Licensor through PLSclear.

Climate change is altering the flammability of temperate ecosystems. Large fires have been linked to higher temperatures that increase the length of the fire season (Westerling et al., 2006), and a warmer atmosphere impacts the availability of fuel to burn due to increased drying (Abatzoglou and Williams, 2016). Particularly in forest ecosystems, climate change–driven drying is already having an impact on forest area burned. Juang et al. (2022) found that, while moisture once governed the speed of fire, as the atmosphere demands more water from an ecosystem (vapor pressure deficit), it pulls moisture from the system. At the same time, climate change is altering fuel availability—specifically, biomass and its availability to burn in an ecosystem—which can lead to larger energy released when burned (Goodwin et al., 2021).

The dry, historically frequent fire systems have a surplus of fuel because of active fire exclusion for over 100 years, combined with past land use and land management decisions, Hurteau explained. Past land use and land management decisions have pushed these historically frequent fire systems past their carbon carrying capacity, from the standpoint of natural processes, such that there is now potential for significant ecosystem change. Climate change is influencing carbon carrying capacity from a productivity standpoint, and the combination of human influences on management and climate are causing fire-prone systems to carry more aboveground biomass than they are capable of. Safford pointed to the accelerating forest loss to wildfire in California. He noted that the focus of destructive fires has shifted from chaparral landscapes in the south to the forests of north and central California, where forest ecosystems are burning at such large scales and high severity that they are not recovering (Safford et al., 2022; Wang et al., 2022). The rapid increase in fire severity in California and the western United States can be explained by variation in fuels, weather, and climate, with fuel accumulation from suppression playing a major role (Safford et al., 2022).

Hurteau explained that achieving a stabilization of carbon would require a reduction in biomass and reintroduction of fire in its appropriate ecological role. While there is a short-term carbon cost to fire restoration work in these ecosystems, dynamic vegetation processes operate on a decadal timescale and large spatial scale. Modeling studies have shown that compared with no management, management actions such as prescribed burns can decrease cumulative fire emissions by modifying the amount of biomass available to burn (Krofcheck et al., 2019). While there is a short-term carbon cost to fire restoration work in these ecosystems, over the long-term these management actions can lead to greater carbon storage by modifying the way fire interacts with vegetation (Krofcheck et al., 2019). Central to restoration of fire in these ecosystems is managing the types and amounts of vegetation and their spatial distribution, Hurteau said.

Arctic/Boreal Biomes

Hélène Genet, University of Alaska Fairbanks, described the ways in which wildfire is a catalyst of change in boreal and Arctic biomes. In the past, wildfire has maintained and promoted ecosystem structure and function across boreal biomes (e.g., Kelly et al., 2013). For example, black spruce forests maintain a thick organic layer to keep soil cool and moist, which promotes low fire severity and low fire return intervals (Johnstone et al., 2011). Stephens provided another example of the Canadian boreal forest in which about 90 percent of ignitions are from lightning, and historically, the ecosystem would have high-severity fires every 40–150 years and could regenerate. On the other hand, evidence from Indigenous knowledge suggests fire severity was highly variable even within boreal fires (e.g., Lewis and Ferguson, 1988).

Today, while wildfires in Arctic and boreal regions represent only 1.2–1.6 percent of global area burned, they contribute 6.3–7 percent of global carbon emissions because fires in Arctic and boreal systems occur mostly in permafrost ecosystems that store large amounts of carbon in the soil, explained Genet (Figure 5) (Veraverbeke et al., 2021). In North America and Eurasia in both forest and tundra regions, belowground carbon stocks dominate emissions (Veraverbeke et al., 2021). Thus, large fires in the boreal can have major effects on the regional carbon balance by releasing carbon that has been stored for decades or centuries (e.g., Genet et al., 2018). After fires burn in permafrost regions, permafrost depth increases, increasing the soil carbon available for decomposition, which can increase post-fire carbon emissions (Gibson et al., 2018).

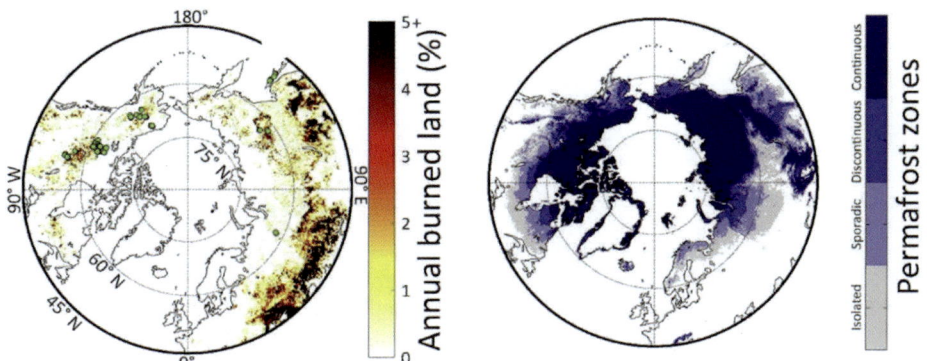

FIGURE 5 Circumpolar maps of annual area burned (left) and permafrost zones (right) for Arctic and boreal biomes for 2001–2020. The Arctic and boreal biomes represent 1.2–1.6 percent (8–10.6 Mha yr^{-1}) of global burned area and 6.3–7 percent (142–210 TgC yr^{-1}) of global carbon emissions. Permafrost (right) is colored by zone in which 42.2 percent of the Arctic/boreal burned area occurs in continuous permafrost terrain, 17.3 percent in discontinuous permafrost terrain, 36.9 percent in sporadic and isolated permafrost landscapes, and 3.6 percent in landscapes without permafrost. SOURCE: Veraverbeke et al. (2021).

Boreal biomes are warming at least twice as fast as the rest of the world, which increases fire frequency and fire occurrence in the boreal and Arctic (e.g., Kharuk et al., 2022). Changes in the fire regime disrupt the legacy cycle of these ecosystems in which increased-severity wildfires burn more of the system's thick organic layer, creating an environment for deciduous tree species to dominate the post-fire succession on the landscape.

This shift to deciduous trees, which have more aboveground biomass, can have consequences for the carbon and nitrogen cycle (Melvin et al., 2015) and dampen the fire regime over the medium to long term (Bernier et al., 2016). Increases in fire severity can also accelerate permafrost loss that may not recover even 100 years post-fire (Jafarov et al., 2013). The increases in fire frequency and fire severity will burn the ecosystem's organic layer, meaning forests will not have enough time to recover the carbon lost during the fire and will begin to lose legacy carbon (Walker et al., 2019).

Stephens pointed to an example in Wood Buffalo National Park in Canada where boreal forests are reburning at frequencies never seen before due to climate change–induced changes to fire regimes. For example, the same area burned in 2004 and again in 2014, killing the regenerating conifer forest and dramatically changing the ecosystem (Figure 6). Canadian forest ecologists and managers are concerned about this potential for type conversion over a large land area, Stephens reported.

FIGURE 6 Frequent burning in boreal forests does not allow enough time for conifers to regenerate, transforming the landscape and its resilience to future fire. Images are from the Wood Buffalo National Park in Northwest Territories, Canada, where boreal Jack Pine "skeletons" (left) and native grass and shrubs (right) show landscape transitions driven by altered fire regimes. SOURCE: Scott Stephens presentation.

Management in the boreal is challenging because boreal forests are adapted to infrequent, high-severity fire. Prescribed burning would kill trees in these forests because they have thin bark and do not resist fire well. One option that Stephens discussed is to increase suppression resources in the boreal regions to reduce fire frequency, but he thought that this would not be successful. Genet mentioned another management option in the managed boreal is to promote more deciduous forest across the landscape to slow the fire regime. However, such a solution would only be successful if the forest and timber sectors are able to adapt and have a market for boreal hardwood.

Genet also identified several potential directions for future research on wildfire in Arctic and boreal biomes: evaluation of long-term effects of wildfire on vegetation, permafrost, and ecosystem carbon dynamics across the boreal and Arctic biomes; representation of drainage conditions and stand characteristics when estimating fire emissions; and representation of ecological shifts associated with changes in wildfire regimes in process-based biogeochemical models. Additional ideas for improved understanding discussed during breakout sessions are summarized in Box 2.

BOX 2
Workshop Participant Ideas to Improve Understanding of Fire in Presently Vulnerable Biomes

Breakout discussions on the first day of the workshop discussed the following: gaps in knowledge of greenhouse gas emissions from fires, uncertainties in fire activity and the carbon budget, and related research areas that could improve understanding:

- Generating data maps of global carbon stocks;
- Improving carbon accounting to include avoided emissions and better classification of emissions as anthropogenic or nonanthropogenic;
- Reducing uncertainties in fuels at the regional, continental, and global scales as well as spatiotemporal variability;
- Improving understanding of post-fire carbon stock recovery over time;
- Improving understanding of feedbacks, including emissions from permafrost thaw, cooling from aerosols, and methane emissions from smoldering emissions;
- Developing a next-generation fuel monitoring system and vegetation mapping, including the three-dimensional structure of fuels; and
- Creating early warning prediction systems that take into account climate and fire weather to predict how much fuel would burn.

This box provides a summary of the breakout discussion. It should not be construed as reflecting consensus or endorsement by the committee, the workshop participants, or the National Academies of Sciences, Engineering, and Medicine.

Tropical Biomes

Susan Page, University of Leicester, introduced the importance of tropical peatlands for carbon storage and the current vulnerability of peatlands to fire. Roughly one-sixth of global carbon stored in peatlands (100 billion tons) is located in the tropics (Dargie et al., 2017; Page et al., 2011). Tropical peatlands are highly vulnerable carbon pools due to multiple large-scale changes: rapid land use change, including agricultural land conversion; use of fire as a land clearance tool with the potential to release both above- and belowground fuel loads; introduction of people to ecosystems and resulting intentional or unintentional human ignition of fires; and climate drivers. As an example of land use changes in Southeast Asia, as of 2015, about 50 percent of peatland was occupied by smallholder agricultural areas or industrial plantations, while only 6 percent remained as forest (Miettinen et al., 2016, 2017).

In 1997–1998, during the strongest recorded El Niño and Indian Ocean Dipole in the 20th century, large wildfires burned in Indonesia and spread haze across Southeast Asia. While this event first drew attention to fires on tropical peatlands, peatland fires have become a recurring issue, often (but not always) associated with El Niño events, explained Page. The GHG emissions associated with these fires are significant, for example, exceeding the fossil fuel emissions of Indonesia in 2015.

Climate change has both direct and indirect effects on peatland fire dynamics. Drought is an example of a direct effect in which water tables in drained peatlands will become even lower, exposing a higher fuel load for combustion. Indirect effects include the combination of land use and climate-driven changes that degrade forests and increase fuel availability. There have also been dramatic changes from people using fire purposefully or unintentionally—as a result of land use change—on landscapes that were not previously flammable.

Peat fires are smoldering fires; they are slow moving, low temperature, can persist over time, and are often in remote locations making them difficult to control (Hu et al., 2018). One approach to calculating total fire emissions from peatlands is the conventional burned-area approach (van der Werf et al., 2010). However, estimating emissions from peatland fires is challenging because there are large uncertainties in burned area, peat density, burn depth, and burn heterogeneity, as well as few measurements of emission factors. A "top-down" approach using, for example, radiative fire energy from satellite observations (e.g., Kaiser et al., 2012; Wooster et al., 2005), also presents challenges for quantifying emissions in tropical peatlands, particularly due to limited data availability and few measurements of emissions.

Page summarized the complex landscape of drivers and impacts of peatland fires in remote, tropical regions (Figure 7). To understand the scale and type of emissions, better understanding of the drivers of peatland degradation (e.g., drainage, clearing, conversion, fire history) and peatland characteristics (e.g., chemical, physical, biological) are important.

There is a long way to go to restoring peatlands and reducing the scale of emissions from wildfires. While there have been efforts to restore some degraded peatlands, logging and other activities have introduced drainage canals that are challenging to block, and it is difficult to keep ignition sources out of the landscape. To make progress, successful rewetting activities—deliberate actions that aim to restore the water table of a drained peatland—together with continuous fire monitoring and fire management are needed, said Page.

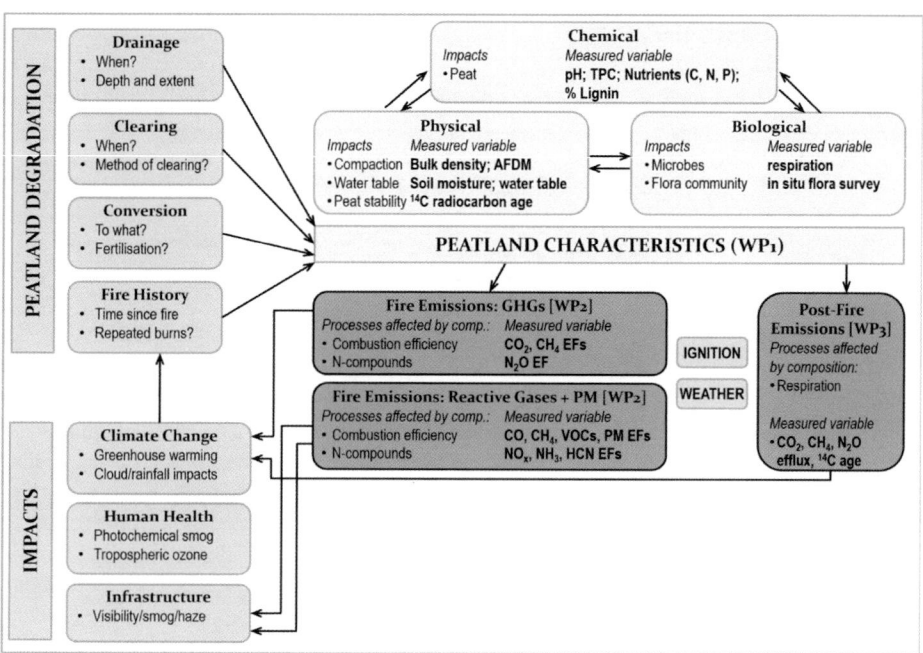

FIGURE 7 Complex landscape of factors that influence peatland fire emissions and impacts. To estimate greenhouse gas (GHG) from peatland fires, information about peatland characteristics (chemical, physical, and biological variables that can be measured) and peatland degradation (drainage, clearing, conversion, and fire history) are important. Emissions from peatland fires also have impacts on climate, human health, and infrastructure. SOURCE: Tom Smith.

Breakout discussions on the first day of the workshop included other important ecosystems that had not been the focus of the workshop thus far. Discussions identified, in particular, nonforested systems (e.g., grasslands, rangelands, woodlands) and other vulnerable ecosystems, including forest clearing and floodplains in the Amazon, Congo, and Sierra Nevada (where there is weakened ecosystem resilience); Mediterranean (where there has been land abandonment); woodlands of India and Southeast Asia; and African savannas, particularly interactions between agriculture and woodland savannas.

MANAGEMENT OF FIRES AND ECOSYSTEMS AND IMPLICATIONS FOR GREENHOUSE GAS EMISSIONS: RECENT PAST AND CURRENT

The final session of the first day focused on present-day management and stewardship approaches for wildland fires and ecosystems and implications of these approaches for GHG emissions. Speakers in this session represented snapshots of practitioner and Indigenous knowledge across different ecosystems and biomes. They described current understanding of the solution space for land stewardship; stewardship approaches that are and are not working in different biomes; the ways that settler and Indigenous approaches to management have influenced resilience; and the short- and long-term benefits of fire management.

Management in Temperate Biomes

Building on the foundational context for the vulnerability of temperate forests in the western United States, described above, speakers outlined a number of strategies to reverse the decades-long policy of fire suppression that has left these ecosystems vulnerable. Safford argued that instead of focusing on burned area as the metric of successful management in these systems, attention should instead be on the conditions under which fires are burning. In California, fuels in fire-suppressed forests need to be managed to slow or reverse current trends, said Safford. Mechanical fuel reduction, or thinning, is challenging because most federal land in the western United States—45.9 percent of the land in the 11 coterminous western states is federally owned—is too steep, distant, or protected to permit mechanical operations by heavy forestry equipment. Additionally, markets for nontimber forest products generated by thinning are underdeveloped. Hand thinning, in which smaller handheld tools are used on smaller diameter materials, can be an effective add-on after mechanical thinning or before prescribed burning, however, hand thinning generates woody residue. While moving fine, small-diameter fuels offsite to be repurposed is an important goal, the sawmill and other wood products industries, particularly in the western United States, have collapsed or been underdeveloped, so there is not sufficient capacity to utilize these fuels.

In the western United States, Safford said, fire itself will be the arbiter of most fuel treatment. Proactive management of how fire operates on the landscape on a large scale can reduce fire risk and increase ecosystem and community resilience. In the absence of pre-fire management, most federal land will be burned by wildfires under current management practices. Proactively allowing fires occurring under benign conditions to burn, instead of extinguishing them, is the only treatment option for 60–70 percent of the federal forested land, Safford explained.

Stephens shared a success story in the Illilouette Creek Basin of Yosemite National Park where lightning fire has been allowed to burn unabated for the last 50 years (Boisramé et al., 2017; Collins et al., 2009; Stephens et al., 2007, 2021). This research demonstrated that managed wildfire could restore a formerly fire-suppressed temperate or frequent fire forest to a functioning fire regime with the capacity to self-regulate (Figure 8).

Riley reiterated that fire needs to be a part of successful fuel treatments, including broadcast burning—burning the whole area—or pile burning, which may be less effective. Modeling can be a useful tool for fire managers to better understand the location and size of future fires. Riley provided the example of using a burn probability map together with datasets of carbon estimated from models to quantify projected emissions from litter, duff, downed woody debris, and standing trees.

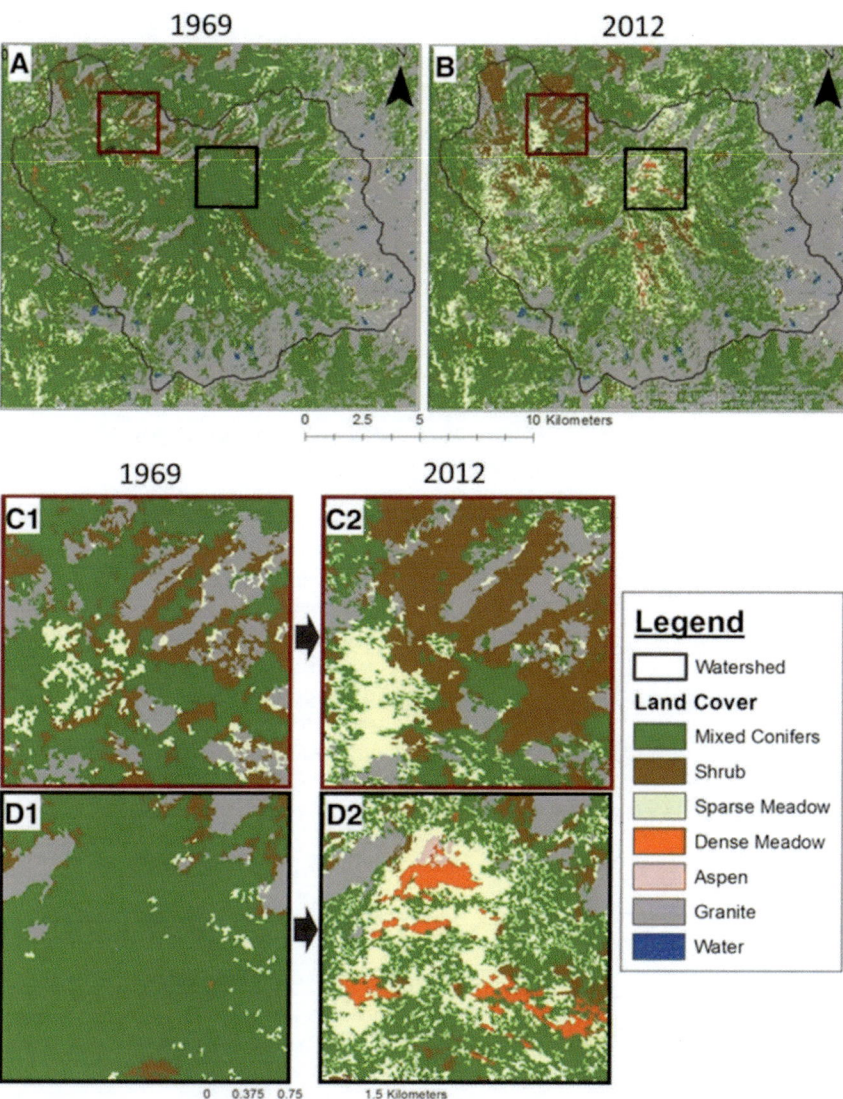

FIGURE 8 Map of land cover in the Illilouette Creek Basin of Yosemite National Park in (a) 1969 (after 100 years of fire suppression) and (b) 2012 (40 years after fire regime change from suppression to lightning fire management). Insets C1 and C2 show an area where both sparse grasslands and shrublands have expanded post-fire, and D1 and D2 show an area that has generally changed from conifer cover to more open vegetation. Over this time period of managed fire, forest cover was reduced by 22 percent, meadow areas increased by 200 percent, and shrublands increased by 24 percent, indicating increased landscape heterogeneity that likely improves resilience to fires. SOURCE: Boisramé et al. (2017).

Stephens offered hope for solutions to reduce fire hazard and promote forest restoration in temperate forests. He emphasized that Indigenous stewardship, which has existed for millennia, has large potential if Indigenous people choose to partner with western scientists (Box 3). Also critical to future management is a workforce that can scale up to work on management year-round, rather than a seasonal workforce built for fire suppression. Stephens noted that in the United States, a true partnership with Indigenous people to manage fire in an ecologically appropriate way is a strategy that has not yet been tried but could be a potential solution. In temperate systems, cultural burning, prescribed burning, and active management of wildfire in appropriate areas can all be done in ways that are complementary to cultural activities.

BOX 3
The Stewardship Project: A Partnership of Indigenous Knowledge and Western Science

Scott Stephens, University of California, Berkeley, and Don Hankins, California State University, Chico, introduced the Stewardship Project,[a] a 50-50 partnership between Indigenous communities and western scientists to produce fire policy recommendations for state and federal governments. Hankins explained that the effort is focused on enabling forest resilience through revitalizing the principles of active stewardship and Indigenous leadership. The project is centered on four key elements: tribal rights to steward, workforce and capacity development, regulatory realignment, and public land fire management areas. This effort is complementary to the Wildland Fire Mitigation and Management Commission that released their report in September 2023 (Wildland Fire Mitigation and Management Commission, 2023), with workforce development being a strong area of overlap, Stephens explained.

FIGURE A culturally informed burn in the blue oak woodlands, Big Chico Creek Ecological Reserve. SOURCE: Don Hankins.

[a] See https://climateandwildfire.org/cwi-projects/the-stewardship-project.

Management in Arctic/Boreal Biomes

Amy Cardinal Christianson, Parks Canada, centered her remarks on the impacts of fire on people in Canada, specifically, on the disproportionate impacts on Indigenous people. Wildfire evacuations are one metric for capturing impacts on people, and Christianson showed that while 5 percent of the Canadian population identifies as Indigenous, 42 percent of wildfire evacuation events in Canada have been of communities with greater than 50 percent Indigenous populations (Figure 9). Communities are evacuated due to both direct threats from fire as well as evacuations solely due to smoke and poor air quality. The same communities have been evacuated, for example, up to seven times in the last 40 years, and almost all these communities with repeat evacuations are Indigenous (Figure 9).

Christianson explained that Indigenous people in the boreal have always "sought to replace fires of chance with fires of choice" (Pyne, 2007) using fire and working with lightning to clean the land to achieve specific cultural objectives. Indigenous practices have been removed from the landscape of the boreal, and this cultural severance has had major implications. The path forward, Christianson argued, should center Indigenous-led fire management in their own territories and dedicated resources to employ Indigenous people in the boreal, rather than primarily partnerships with agencies that want to learn from Indigenous knowledge (Christianson et al., 2022; Maclean et al., 2023). The northern territories of Australia are an example of where Indigenous fire management has reduced emissions.

Reflecting on opportunities to learn from Indigenous practices, Christianson shared that Indigenous people in the boreal work on a circular calendar based on observations of the land around them, which is a practice highly adaptable to climate change. The challenge, however, is how to start reintroducing these practices after the forest has been so mismanaged, where the historical fire practices that Indigenous people used could not be done safely today.

Dan Thompson, Canadian Forest Service, explained that the 2023 fire season in Canada was by far the worst on record nationally, leaving the entire country impacted by severe fire. The 2023 fire season reflected the forest environment in Canada and was largely a consequence of lightning rather than human ignitions. In September 2023, at the time of the workshop, less than 40 percent of fires were being actively suppressed because the fires were too large and too remote.

Forest conversion to deciduous species is one of few management and mitigation strategies feasible on a large forest scale, Thompson said. After fire, young forest, especially broadleaf, is commonplace from resprouting and air-transported seeds; these forests are one of the few natural negative feedbacks to dampen an otherwise accelerating fire regime in the boreal. Young broadleaf is a consequence of natural succession in the Canadian boreal and burns at 12–25 percent of the frequency (Bernier et al., 2016), has lower landscape-level flammability (Erni et al., 2018), and directly emits approximately one-third the direct emissions of smoke and GHGs, compared to the Canadian forest as a whole. Fire propagation modeling suggests that stitching together recent wildfire scars with prescribed burning, particularly in areas changing from spruce to less flammable broadleaf, can start to have landscape effects that could impact fire probability, even within old conifer forests. Thompson also noted that some prime areas where there has been the most fire suppression over the last 75 years is in the roughly 40 kilometers around dispersed communities in northern Canada that could, at the local scale, be a place to start to reintroduce fire and conversion to a less flammable landscape.

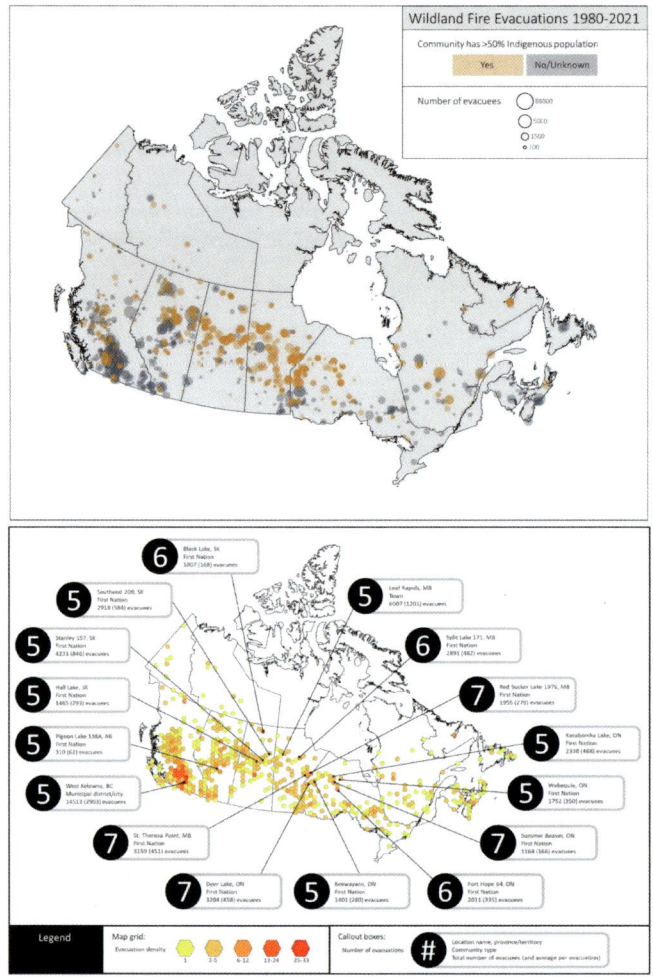

FIGURE 9 (Top) wildfire evacuations in Canada from 1980 to 2021. Colors indicate where communities are over 50 percent Indigenous (orange), and the size of the circles shows the number of evacuees. (Bottom) Evacuation frequency where callout boxes show examples of the number of repeat evacuations by community type across Canada and colors show evacuation density where yellow is low (1) and red is high (25–33). SOURCE: Amy Cardinal Christianson presentation.

Management in Tropical Biomes

Bibiana Bilbao, Universidad Simón Bolívar, described the steady increase in active fires across South America over the last decade in areas where there is land use change from deforestation and hotter and drier climates (Anderson et al., 2022). In response to extreme conditions and the occurrence of megafires—for example, in the Amazon—most countries in South America have adopted zero-fire policies that focus on avoiding fire and directing resources toward fire brigades and technical support. Not only are these methods not effective, but also the $5 billion per year that the United States spends to fight forest fires is not feasible for countries in Latin America. Further, Bilbao explained that these zero-fire practices come from colonization, excluding local Indigenous knowledge and practices that allowed cultures to survive and preserve a highly diverse continent.

Bilbao shared a new paradigm of integrated fire management with an intercultural vision, drawing on experiences working with the Pemón Indigenous peoples in Canaima National Park, Venezuela, in northern Amazonia. The Pemón use fire in their daily practices for agriculture and hunting, as well as to protect forests from fires that start in savanna areas. Bilbao implemented a collaborative, long-term experiment to simulate traditional Pemón fire methods by burning a series of plots over different time periods (Bilbao et al., 2009, 2010, 2017, 2021, 2022). The technique of creating a mosaic of patches with different fire histories, known as patch mosaic burning (PMB), can be used as a firewall to reduce the risk of hazardous wildfires, particularly in the savanna–humid tropical forest transition (Figure 10). This technique imitates practices used by the Pemón Indigenous Peoples for centuries to sustainably manage savanna–forest boundaries and protect the forests that represent their principal source of food resources and spiritual reasons. If fire is excluded from these savanna areas, dried fuel accumulates, increasing the risk of wildfires spreading into the forest. Bilbao also emphasized the importance of traditional management in the context of climate change to reduce wildfire risk. In addition to incorporating Indigenous knowledge and participation in new intercultural paradigms of fire management, it is equally crucial to honor Indigenous cultural heritage and acknowledge Indigenous peoples' capacity to evolve and adapt their practices to new conditions, Bilbao explained.

Cynthia Fowler, Wofford College, drew attention to the subset of GHG emissions from subsistence economies, known as survival emissions. Fowler shared that for islander communities in the Indian Ocean, landscape burning is essential for managing subsistence regimes, acquiring adequate nutrition, experiencing the world in meaningful ways, expressing cultural identities, and exercising self-determination. While GHG emissions from Indian Ocean nations have been increasing over the past two decades, some of the emissions are generated while meeting basic needs, including from agriculture, animal husbandry, and biomass burning. Survival emissions can be difficult to measure because they are often dispersed, small, informal, and unpublicized.

One example of these systems is the Indian Ocean island, Sumba, where savanna and garden fire environments are managed through pasture livestock, weeding, planting, harvesting, and burning. Sumba has two overlapping anthropogenic fire regimes in which there is fast burning of fine fuels, which produce fewer emissions. Relative to total GHG emissions from agriculture in Indonesia, for example, survival emissions from burning crop residue contributes few emissions while enabling livestock husbandry and food production. Fowler argued that these fire regimes are worth studying to understand the human and ecological conditions where management produces fewer emissions.

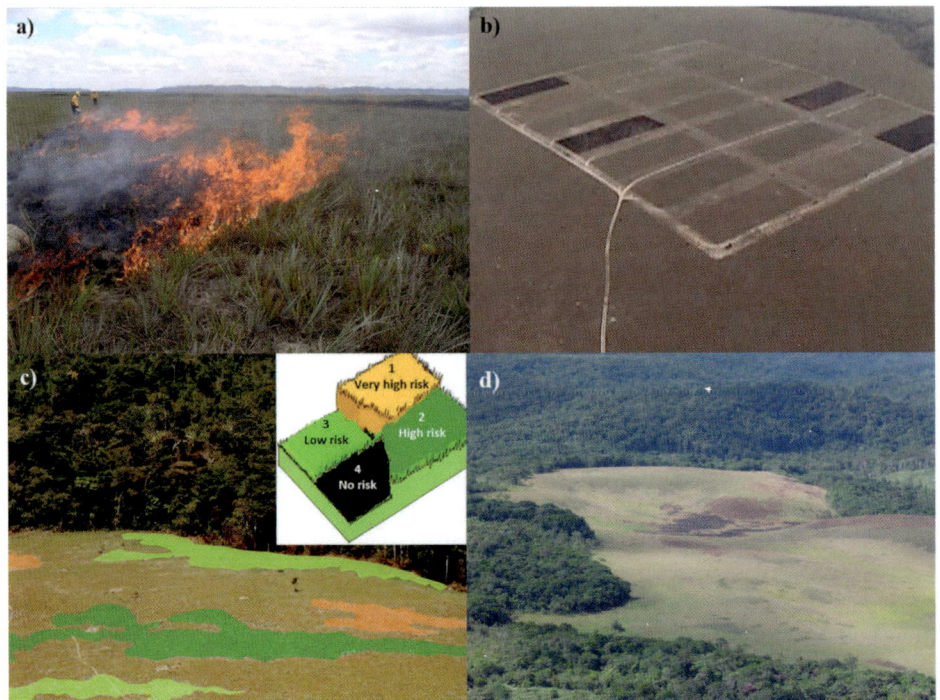

FIGURE 10 (a) Experimental burns initiated by Indigenous Pemón firefighters in Canaima National Park, Venezuela; (b) Aerial view of the long-term fire collaborative Interactions Atmosphere–Biosphere of the Gran Sabana, Bolívar State, Venezuela experiment plots; (c) and (d) Different views of Indigenous patch mosaic burnings. SOURCE: Bilbao et al. (2021).

Fowler pointed to the need for local studies on biomass fires in diverse environments. She argued that assessing the contribution of survival emissions from rural, Indigenous, and agropastoral communities relative to anthropogenic emissions from other communities around the world raises questions about which parts of society are vulnerable to climate change, and which parts are responsible for reducing emissions. Fowler underscored the high stakes for Indigenous islanders and need for their leadership in research projects as well as policy and management recommendations stemming from such studies.

Examples of Additional Management Opportunities from Breakout Discussions

Breakout discussions highlighted additional management opportunities across different biomes. Discussions reiterated workshop themes about the importance of partnering with Indigenous communities who have been stewards of fire on the landscape for millennia. Participants emphasized the importance of considering fire ecology in the local context and history of an area, and considering consequences of management policies on water, biodiversity, and community resilience.

Discussions also highlighted locations with a mix of fire severity—for example, the western United States—as a potential entry point on the landscape for continued management, including prescribed burning and fuels management. Rapid detection tools to target fire suppression in vulnerable ecosystems with rapidly changing fire regimes and deep ancient carbon could be a potential strategy, together with prescribed burning, fuels treatment, and cultural burning. Another discussion highlighted strategic pre-fire planning, or potential operational delineations, to design fire response strategies before fire in collaboration with local interested and affected groups.

Observing and Modeling Wildland Fires and Their Greenhouse Gas Emissions: Opportunities and Challenges

The second day of the workshop focused on the tools used to quantify greenhouse gas (GHG) emissions from wildland fires and better understand the impact of fires on the net carbon budget. The first session focused on observational-based approaches for quantifying emissions and the second session focused on modeling approaches. Both sessions as well as the breakout discussions on the second day sought to identify key challenges and gaps that could improve understanding of GHG emissions from wildland fires today and in the future and inform potential mitigation strategies.

OBSERVATION-BASED APPROACHES FOR QUANTIFYING EMISSIONS FROM WILDLAND FIRES

Bo Zheng, Tsinghua University, described the importance of accurately quantifying emissions from wildland fires to inform mitigation and adaptation policies. Broadly, there are two methods to estimate GHG emissions from fires: (1) activity-based (or "bottom-up") approaches combine activity data about fires with emission factors, and (2) atmospheric-based (or "top-down") approaches use observed atmospheric concentrations together with models (Figure 11). On the one hand, activity-based approaches can quantify multiple species in near real time but may not represent small fires or represent fire dynamics. On the other hand, atmospheric-based approaches are consistent with observed smoke pollution but are limited by which species satellites observe and struggle with constraining CO_2 emissions in particular.

Activity-Based Approaches

Andy Hudak, U.S. Forest Service, dove deeper into the bottom-up or activity-based approach for quantifying emissions from wildfires and highlighted opportunities for ground-based observations to reduce key uncertainties. The activity-based approach estimates fuel consumed either indirectly—based on fuel load estimates pre- and post-fire (Figure 12a)—or directly—based on observations of heat flux as fuel is combusted (Figure 12b) (French and Hudak, 2023). The largest contributors to uncertainties in emission estimates are (1) the process of consumption, which is highly variable; and (2) fuel beds, which are complex, heterogeneous, and challenging to measure on the ground (Larkin et al., 2012; Prichard et al., 2022). Burned-area maps from fire records or remote sensing are relatively better constrained.

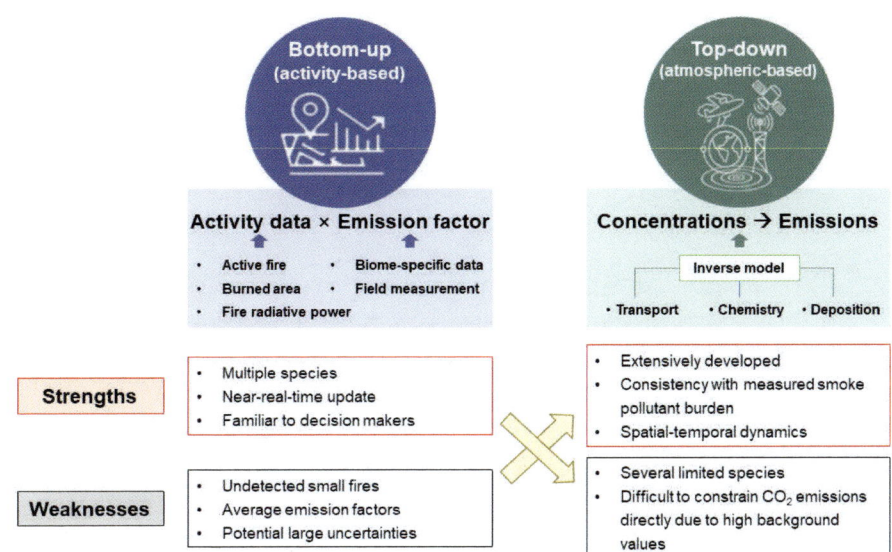

FIGURE 11 Two general approaches for quantifying emissions from wildland fires. (Left) Activity-based, or "bottom-up," approaches combine activity data about fires, such as observations of burned area or fire radiative power, with emission factors. (Right) Atmospheric-based, or "top-down," approaches combine observations of atmospheric concentrations of smoke with inverse models. The bottom panels show the strengths and weaknesses of each approach, and the arrows indicate opportunities for combining the approaches to leverage their strengths and address their weaknesses. SOURCE: Adapted from Bo Zheng presentation.

Combustion accounting depends on many variables, including dead versus live vegetation and the amount of fuel moisture, which is coupled to soil moisture but more temporally dynamic, especially for fine fuel components such as senesced grass and litter that drive surface fire behavior as opposed to coarse woody debris and duff. Quantifying fuel beds is challenging due to their complexity. Remote sensing observations, particularly three-dimensional LiDAR (light detection and ranging) data, can provide information about fuel structure; however, the measurement sensitivity is mainly to the overstory canopy structure rather than the underlying litter and duff. Hudak explained that this is why large amounts of ground data are needed to capture the variability in fuels and burning conditions, which are key to understanding fire ecology, behavior, and effects. Hudak provided an example of using LiDAR measurements of the overstory to constrain estimates of litter inputs to a surface fuel bed, which can inform the ultimate goal of creating heterogeneous fuel maps (Sánchez-López et al., 2023). In addition to ground measurements of fuels, Hudak also noted the need for ground measurements of consumption from prescribed burn experiments. Prescribed burn experiments can provide multiscale data of how fuel types burn under different conditions (e.g., fire weather, fuel moisture) that can also be used to inform model estimates of emissions.

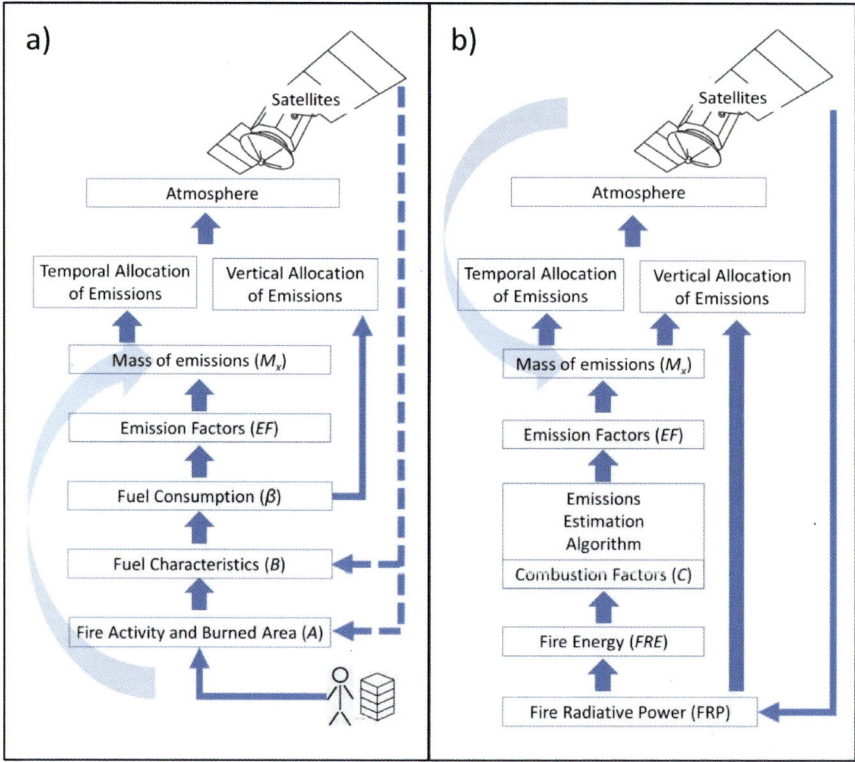

FIGURE 12 Activity-based or bottom-up approaches to quantify emissions from wildland fires. Fuel consumption can be estimated using two approaches: (a) the indirect approach that considers pre-fire fuels compared to fuels consumed; or (b) the direct approach that uses observations of fire radiative power to estimate emissions based on fuel-specific combustion factors. SOURCE: French and Hudak (2023).

Hudak explained that the indirect approach using pre- and post-fire measurements has the disadvantage of not providing information about how fuel is consumed between the two points in time. Using a direct approach, such as heat flux during a fire, allows for estimates of how much combustion was flaming versus smoldering over the course of a fire—information needed to quantify the resulting emissions.

Jeff Vukovich, U.S. Environmental Protection Agency (EPA), described the bottom-up approach of the National Emissions Inventory (NEI) that quantifies emissions from anthropogenic sources, biogenic sources, and fires in the United States every 3 years. The NEI produces a day-specific emissions inventory of air pollutants, methane, and carbon dioxide (CO_2) for prescribed burns and wildfires. EPA receives activity data from states, localities, tribes, and federal agencies on a voluntary basis and combines that information with satellite observations from the National Oceanic and Atmospheric Administration's Hazard Mapping System.[1] To generate the inventory of emissions, the Smartfire2 tool is used to reconcile data sources and produce daily acres burned by fire type, and the U.S. Forest Service's tool, Bluesky Pipeline, is used to estimate emissions for both flaming and smoldering phases (EPA, 2023). The NEI fire emissions are typically used for emission trend reporting, regulatory air quality and risk assessment modeling, and in research (e.g., air quality modeling). The Inventory of U.S. Greenhouse Gas Emissions and Sinks is a separate inventory produced at a national resolution each year.[2] Vukovich outlined a number of gaps and uncertainties in the NEI wildfire emissions, including activity data for prescribed fires, plume rise algorithms, methods to include pile burns, and emission factors. Vukovich also pointed to the opportunity to put more resources toward quantifying emissions in Alaska, particularly in non-NEI years when the fire activity may be high.

Quantifying Emissions from Satellite Observations

Louis Giglio, University of Maryland, discussed opportunities, limitations, and uncertainties for using satellite observations to understand wildland fire emissions. Satellites can observe many different parameters that can inform wildfire emission estimates, including

- Active fire presence: one or more fires actively burning within the footprint of a sensor;
- Fire radiative power (FRP)[3]: mid-infrared data show the presence of fires;
- Burned area: mapping cumulative area burned retrospectively;
- Land cover: classification of land cover types used as indicators of fuel type, or high-resolution fuel-specific maps; and
- Physical and meteorological parameters (e.g., fuel moisture, soil moisture, precipitation, etc.).

Giglio explained that satellite observations are the only practical way to comprehensively observe fire activity over the global biosphere. In general, the quality of satellite observations is improving and the data are generally freely available. However, the satellite data record is short compared to climate or human timescales. Additionally, the quality and consistency of satellite data vary over time and space, and the data are often stretched beyond their intended purpose to estimate wildfire emissions. Specifically, satellite observations of active fires and burned area can have omission (i.e., missed fire or burn) and commission (i.e., false fire or burn) errors, FRP observations can have errors in the analytical approximations used, and land cover classes can be misclassified.

---

[1] See https://www.ospo.noaa.gov/Products/land/hms.html.
[2] See https://www.epa.gov/ghgemissions/inventory-us-greenhouse-gas-emissions-and-sinks.
[3] FRP provides an instantaneous burn rate, and fire radiative energy (FRE) is the time-integrated product of FRP and can provide the total biomass burned by a fire.

These uncertainties can all influence estimates of emissions; specifically, emission estimates can be overestimated or underestimated, systematically biased, assigned to the wrong time or place, or incorrectly attributed to fire or the wrong species. Further, uncertainties also depend on scale. Giglio identified the largest gaps or uncertainties facing the use of satellite data for quantifying fire emissions:

- Small burns that may be missed due to low spatial and temporal resolution,
- Limited diurnal sampling from polar orbiters,
- Less sensitive or spatially comprehensive sampling from geostationary sensors,
- Discrepancies in land cover datasets (e.g., Zubkova et al., 2023),
- Lack of long-term continuity and consistency across sensors and orbits, and
- Attention to quality assurance and validation.

Regarding key satellite observations needed to identify strategies to reduce GHG emissions from wildfires, Giglio said that forest and peatland fires are the most important from an emissions perspective, and that new and planned geostationary satellites, supplemented with polar orbiters at high latitudes, hold the most promise for observing these emissions with high frequency.

Drilling down into one example satellite observation, **Martin Wooster, King's College London,** detailed the approach of quantifying wildfire emissions from FRP observations. The FRP approach for estimating fire emissions is the most direct approach that requires the least assumptions and can produce estimates of fire emissions in near real time, Wooster said. Imaging in the mid-wave infrared is sensitive to high-temperature sources, even if they are only a small part of a satellite pixel. Analysis of mid-infrared data allows for the presence of fires to be detected, including small fires observed at low spatial resolution. Observations are processed by active fire detection algorithms developed to be sensitive to the smallest possible fires while being insensitive to signals that are not fires. Generally, detectability scales with pixel area—in other words, if satellite pixels are 10 times larger in area, the minimum FRP that can be detected is 10 times larger. The FRP approach is considered a direct approach because ground measurements have shown that FRP is linearly related to the rate of fuel consumption. In practice, FRP data are converted into wildfire emissions using either a fixed conversion factor or derived relationships between fire radiative energy (FRE) and fuel consumption or smoke plume data (e.g., Kaiser et al., 2012; Nguyen et al., 2023).

Key uncertainties and challenges for the FRP approach include detecting cooler (e.g., tropical peat) fires, validation challenges due to rapid changes in FRP, atmospheric correction of observations, and conversion approaches of FRP to fuel consumption and emissions. Wooster sees opportunities in high-resolution geostationary satellites that can provide continuous observations. Regarding optimal spatial resolution for estimating emissions, Wooster's work has found that 200-meter resolution is sufficient (Sperling et al., 2020).

Giglio explained that between FRE-based and inventory-based approaches, FRE-based approaches avoid the need for detailed information about fuels but require high-frequency temporal sampling; inventory-based approaches require detailed information to estimate what fuel burned, but do not require as high a resolution for temporal information.

Hybrid Approach for Quantifying Emissions

Zheng showed a hybrid approach for quantifying emissions that combines both activity-based and atmospheric-based approaches to estimate carbon monoxide (CO) and CO_2 emissions from fires. In Zheng's approach, he uses satellite observations of CO—an important tracer for smoke from wildfires—to constrain CO_2 emissions using an inverse model (Zheng et al., 2018a, b). Then, he uses CO emission factors to derive combustion efficiency, which determines how much carbon, or CO_2, is released from fires globally (Zheng et al., 2021). Using this approach, Zheng has found that globally, burned areas have decreased since 2000 while fire emissions per unit area burned have slightly increased (Zheng et al., 2021). This hybrid approach allows for investigation of these trends; the decrease in burned area globally was due to declines in grassland burning while emissions from forests have increased, primarily in boreal regions where emissions increase when there are high water deficits (Figure 13) (Zheng et al., 2023).

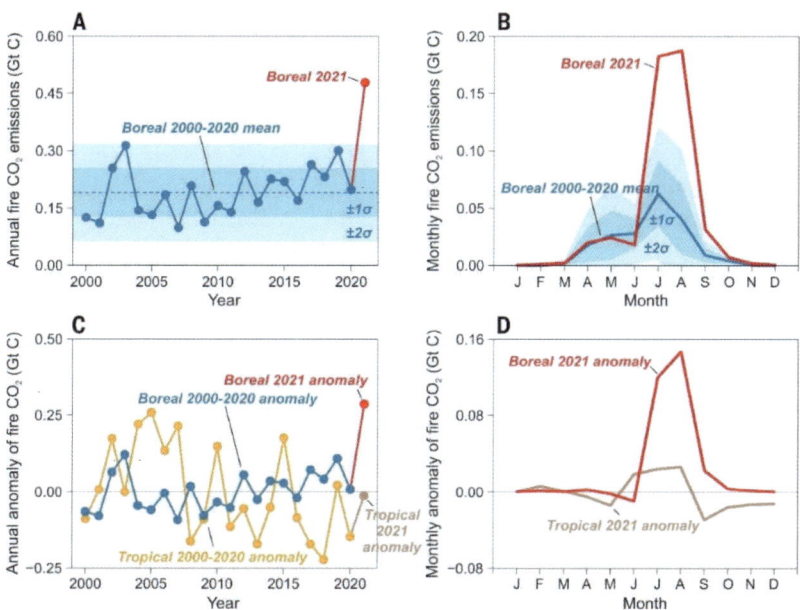

FIGURE 13 Estimates of boreal and tropical fire carbon emissions from 2000 to 2021. Boreal fire (A) annual and (B) monthly emissions with 1 and 2 standard deviations (σ) from the 2000 to 2020 mean shown in blue shading. The (C) annual and (D) monthly anomalies compare boreal emission anomalies with tropical emission anomalies in 2021. SOURCE: Zheng et al. (2023). Reprinted with permission from AAAS.

Challenges in Quantifying Emissions from Prescribed Fires

Morgan Varner, Tall Timbers, focused on key uncertainties and challenges in quantifying emissions from prescribed fires. Prescribed fire is not monolithic; it includes a diversity of fuels, fire behavior, and emissions. In the United States, the largest area where prescribed burning happens annually is in the Southeast, where it is used primarily for biodiversity conservation and wildlife. Varner noted that compared to wildfires, prescribed fires are underrepresented in terms of research funding and attention in scientific publications. Varner reiterated that key uncertainties for understanding emissions from prescribed fires are the spatiotemporal heterogeneity of fuels, fire behavior, and resulting fuel consumption. Estimating the quantity, location, and timing of prescribed burning in the United States is challenging, whether the data are coming from state forestry agencies or satellites (e.g., Cummins et al., 2023; Melvin, 2022; Vanderhoof et al., 2021). For example, permit data for prescribed fires in the United States are an imperfect metric. Permits only capture an estimate of the acres that will be burned; these data are not updated based on how much was actually burned, and the data collected vary state to state. Regarding the use of satellite data to quantify emissions, it is difficult to attribute observations to prescribed burning versus wildfires or agricultural burns.

Varner highlighted the importance of scale for detecting variations in consumption and fire effects, including the heterogeneity at the stand scale (Kreye et al., 2020). There are large uncertainties in characterizing fuels in fine-fuel-driven systems, including for solid fuels such as duff and peat. For example, although duff is an important generator of emissions, it is spatially patchy, and more work is needed to improve predictive models, Varner said. There are also large uncertainties in prescribed fire behavior that have complex, intricate ignition patterns unlike large wildfires. Breakout discussions also noted the lack of data on the effects of prescribed fire on subsequent fires (Box 4).

MODELING EMISSIONS FROM WILDLAND FIRES

The second session of day 2 provided an overview of different approaches for modeling emissions from wildland fires. Specifically, speakers focused on simulating fire in global dynamic vegetation models and lessons learned from the Fire Modeling Intercomparison Project (FireMIP), modeling forest fire interactions in the western United States, statistical methods for defining global pyrogeography and pyromes, and regional-to-global projections of fire emissions.

Global Fire Models

Stijn Hantson, Universidad del Rosario, introduced global fire modeling and presented several example applications. Fire models aim to represent the main drivers of fire occurrence, including the climate, vegetation, and human drivers (Figure 14). A key consideration for models is how they represent the interactions between the various drivers.

> **BOX 4**
> **Examples of Data Needs to Improve Estimates of Wildfire Emissions and Inform Management Strategies**
>
> Breakout discussions on the second day of the workshop focused on a range of gaps in data and observations that could improve emission estimates. Discussions underscored the importance of improving maps of land cover and fuels at consistent and appropriate spatial scales, including at the belowground level and for coarse woody debris. Improved maps combined with more combustion measurements from representative biomes and pyromes could enable understanding of the carbon stock consequences of management actions. Another breakout discussed the gap in fuels information by class (rather than total) to better understand how the fuel load impacts fire behavior and greenhouse gas (GHG) emissions, including for peat and belowground carbon stocks.
>
> Several participants also highlighted opportunities to improve satellite monitoring systems to map carbon stocks and fire emissions through measurements optimized for fire and vegetation with finer spatial resolution and better diurnal coverage. Other discussions identified the gap in site-level data, due to the lack of global capacity, which could support calibration and validation of both models and satellite observations. These sparse data have the consequence of making models biased, for both current and past datasets.
>
> Breakout discussions also explored the challenge of quantifying the impacts from different management strategies—including prescribed burning, thinning, and land use—on fuels, fires, and future emissions across all at-risk ecosystem types and biomes. To quantitatively connect management decisions and GHG emissions, some participants suggested using field and LiDAR data on fuel loads to quantify the resulting emissions from different strategies.
>
> *This box provides a summary of the breakout discussion. It should not be construed as reflecting consensus or endorsement by the committee, the workshop participants, or the National Academies of Sciences, Engineering, and Medicine.*

For example, if a fire model is not coupled to a vegetation model, it cannot capture the interactions and feedbacks between climate change and vegetation status or the feedbacks of fires on vegetation. Dynamic global vegetation models (DGVMs) are critical for incorporating these processes into global fire models, Hantson said. DGVMs simulate ecosystem processes based on a simple set of input parameters (e.g., sunlight, CO_2, soil properties) (Prentice et al., 2007). Fire influences all of these ecosystem characteristics, for example, tree mortality and competition between species, among others. DGVMs provide fuel loads and ecosystem status as dynamic inputs to fire models.

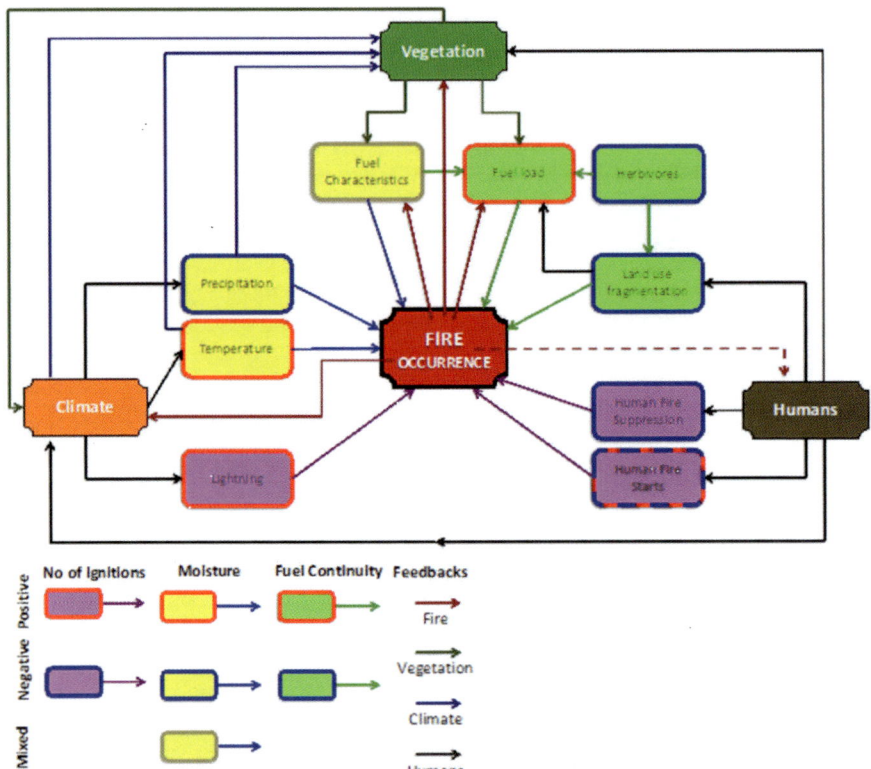

FIGURE 14 Summary of the interactions between the controls on fire occurrence on coarse scales in global fire models. Green boxes represent controls influencing fuel, yellow boxes represent controls influencing moisture, and purple boxes represent controls influencing ignition. Influence on fire is indicated by box outlines as positive (red), negative (blue), or mixed (brown). Arrows indicate interactions between people and other controls (brown), interactions between vegetation and other controls (dark green), feedbacks from climate (dark blue), direct effects (black), and feedbacks from fire (red). SOURCE: Hantson et al. (2016).

Hantson described two main categories of global fire models. One type is the simpler empirical global fire model that simulates a global pattern of burned area based on empirical equations with inputs such as climate drivers, fuel characteristics, and impacts of humans. The second type is the process-based global fire model that simulates the chance that ignition will turn into fire, and when an individual fire occurs, how fast it spreads, and how much area is burned. This individual fire behavior is then upscaled to represent a global pattern of burned area. There is a range of global fire models that have large variations in their inputs and parameterizations and which processes are accounted for, but all models produce estimates of burned area (Rabin et al., 2017).

FireMIP grew out of the need to methodically evaluate the capabilities and uncertainties of existing global fire models (Hantson et al., 2016; Rabin et al., 2017). FireMIP was a community effort to analyze and benchmark the existing global fire models and systematically understand and decrease uncertainty in global fire model projections. Comparing how models simulate global burned area and fire emissions, models generally reproduce the global pattern, with most burning across tropical savannas (Hantson et al., 2020). Seasonality in burned area and fire carbon emissions is also generally well represented by the models, with peak burning better represented than the length of the fire season (Hantson et al., 2020). Hantson also showed a study that used a global fire model to compare burned area changes under climate change from reanalysis data with another dataset without climate change and they were able to show that climate change has increased present-day global burned area by 16 percent (Burton et al., 2023).

The DGVMs generally also reproduce observed relationships with climate variables, but models have trouble representing the impact of changes in previous-season plant productivity and increasing burned area, which may explain why models struggle to represent interannual variability (Forkel et al., 2019). As one example of how limitations in vegetation models impact global fire models, vegetation models have a hard time representing soil organic layers, which if not simulated, cannot then be combusted in a global fire model and may lead to large underestimates in carbon emissions, for example, in boreal North America.

Approaches for Understanding Regional Fires

Park Williams, University of California, Los Angeles, focused on modeling forest fires in the western United States. One of the main drivers of the increase in annual area burned in forests is the warming and drying of the atmosphere. Vapor pressure deficit—the difference between the amount of moisture in the air and how much moisture the air can hold when it is saturated—has increased with increasing temperature, and there is a strong exponential relationship between vapor pressure deficit and burned area (Juang et al., 2022). However, Williams explained that the future will be more complicated than simply extrapolating this historical exponential relationship.

Williams showed results from a new statistical wildfire model for the western United States that responds to changes in daily climate variables (e.g., vapor pressure deficit, precipitation), ecosystem variables (e.g., forest biomass), and human variables (e.g., distance from populations) and generally represents observations of regional and seasonal variations in area burned. When their statistical model is applied to the historical and future period without coupling between the fire and dynamical vegetation model, the statistical model predicts an explosion in area burned by the end of the century. However, ecosystems are critical for regulating area burned; forest fires can have a self-regulating effect in which subsequent fires do not burn as large or intensely through an area that has burned recently. At the subcontinent scale in the western United States, the degree to which annual area burned will increase in the future may be strongly influenced by the self-regulating impact of forest fires (Abatzoglou et al., 2021; Turco et al., 2023).

Complex models are needed to make future predictions about forest fires in the western United States, explained Williams. A mid-range (i.e., 1 kilometer) forest ecosystem model—for example, Dynamic Temperate and Boreal Fire and Forest-Ecosystem Simulator (DYNAFFOREST)—is an example of a compromise between landscape-scale and global-scale models (Hansen et al., 2022). When Williams couples his statistical fire model with the DYNAFFOREST ecosystem model, by the end of the century, while fires are still large, they are about 15 to 20 percent smaller on average due to this self-regulating impact, compared to the uncoupled model. Beyond area burned, this self-regulation also leads to approximately 20 percent less carbon burned per area compared to the 20th-century mean.

Regional fire modeling is relatively nascent, and Williams identified a number of major limitations:

- Simple representations of fuels do not yet include the effects of ladder fuels, ecosystem types other than forests, or insects and diseases;
- Effects of prior or subsequent burning (i.e., self-regulating behavior) are difficult to validate;
- Observations of biomass combusted to validate regional models are limited;
- Simulations of alternative management approaches are challenging because models have been developed on data from the suppression era;
- Uncertainty exists in the effects of CO_2 fertilization on forests;
- Expertise across disciplines can be beneficial to partner in large teams; and
- Computational resources are important to support the coupling and high resolutions for future projections.

Williams argued that modeling wildfire at the regional and continental scale would require a large investment in resources, similar to historical investments in global climate modeling.

Matthew Jones, University of East Anglia, showed new work on regional mapping that reveals shifts in the global geography of forest fires. Archibald et al. (2013) applied clustering analysis to observations of fire characteristics in order to group different parts of the world with similar fire characteristics, defined as global pyromes. These pyromes are useful for studying fire regimes, but they do not provide information about what these pyromes are sensitive to—for example, climate change, land use change, or changes in vegetation productivity.

To address these limitations, Jones introduced "pyronia"[4]—regions with similar fire controls, including climatic factors (e.g., soil moisture, fire weather index), vegetation factors (e.g., fuel stocks), and human factors (e.g., population density, cropland cover). Clustering analysis—a machine learning approach—was used to identify regions where forest burned area has similar sensitivities to the wide range of potential controls on fire. Machine learning techniques can be a "closed box" that requires disentangling to understand and

---

[4] Note: Pyronia is also a genus of butterflies; in context of this proceedings, pyronia refers to regions with similar fire controls.

interpret the results. Jones showed how the distribution of correlations across ecoregions within each pyronia can provide insights about what the clustering algorithm identified as unique about each pyronia. For example, Jones's analysis showed that extratropical pyronia in the temperate and boreal forests are sensitive to fire weather and/or drought and are not sensitive to human controls (e.g., land use, human ignitions).

Once these pyronia have been identified, they can be used to explore how fire behavior has changed in recent decades. Jones showed an increase in burned area, fire severity, and ultimately carbon emissions in the extratropical and suppression zone (i.e., regions where fire suppression has been a dominant forest management practice) pyronia, and decreasing burned area, severity, and carbon emissions in the tropical and subtropical forests. Jones noted that while this approach did not utilize a global fire model, there are ways the two modeling approaches could complement one another—for example, global model parameterizations could be defined or tuned to each pyronium.

Projecting Future Fire

There is growing interest in making projections of future fire to understand a range of risks including meteorological risk, burned area, and carbon emissions. In a comparison of Earth system models that simulate fire, Kloster and Lasslop (2017) showed that there is wide variation in how these models simulated burned area in the late 20th to early 21st centuries, both compared to each other and against observations. **Jed Kaplan, University of Calgary,** argued that global fire models are not successfully representing fire and that more work is needed in this space. Global fire models show large variations in representations of carbon emissions in the past, a convergence in the late 20th and early 21st century due to model tuning to independent estimates of fire emissions, and divergence in their projections of fire emissions in the future (Figure 15) (Kloster and Lasslop, 2017).

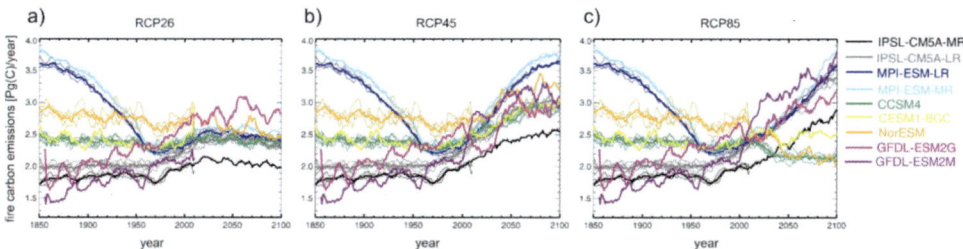

FIGURE 15 Modeled global fire carbon (total carbon; C) emissions from 1850 to 2100 from nine Earth system models (colored lines) that participated in the 5th Coupled Model Intercomparison Project and reported burned area and/or fire carbon emissions. Panels show results under different representative concentration pathways (RCPs). SOURCE: Kloster and Lasslop (2017).

Kaplan elaborated on the differences between model projections of the future by showing experiments that used one fire vegetation model together with different global climate models and future emission scenarios. Climate model simulations of burned area and fire carbon emissions in both the recent past and end of century are different, indicating large uncertainties (Koch and Kaplan, 2022). Examining the spread of models shows that while there are large differences in the magnitude of carbon emissions depending on which meteorology the fire model uses—indicating a model bias—the models generally simulate similar trends, mainly because the models are driven by the same GHG emission scenarios in which there is a direct effect of CO_2 on vegetation and an indirect effect of CO_2 on climate (Koch and Kaplan, 2022).

Kaplan outlined major drivers of uncertainty for projecting future fire: (1) simulated meteorology, (2) fuels and their structure, (3) spatial resolution and landscape heterogeneity, and (4) "anthropogenic fire" or how people use fire today. Regarding uncertainties in anthropogenic fire, Kaplan pointed to the Global Fire Use Survey, an international effort to map the ways humans interact with fire on the present-day landscape. Regarding gaps in measurements, Kaplan reiterated comments from other speakers on the need for data on fuels, particularly in developing countries; future land use projections, particularly related to fuels management and anthropogenic fire; and observations of small fires, particularly as part of smallholder land management. Additional opportunities to improve model estimate of GHG emissions from fires are noted in Box 5.

BOX 5
Opportunities to Improve Model Estimates of Emissions from Wildland Fires

Speakers discussed key uncertainties that, if addressed, could improve global estimates of total fire emissions. Jed Kaplan pointed to opportunities to improve model sensitivity to carbon dioxide (CO_2), for example, simulating the ways global vegetation will respond to CO_2 by the end of the century. Park Williams agreed and added that accurately simulating hydrology at a fine scale, for example, live fuel moisture, could improve the predictability of burning probability. One breakout discussion also noted the importance of understanding water availability for biomass production and fuels, for example, the interactions between wetlands, climate change, and implications for fire behavior.

From Stijn Hantson's perspective, to improve future projections specifically, understanding the limiting factors around why fires get extinguished would improve process-based representations of fires in global models. Breakout discussions highlighted the challenge of integrating human-driven actions, including land stewardship practices, into models to make future projections. Participants also pointed to the importance of simulating new extremes in fire behavior, greenhouse gas emissions, and long-term consequences of fires on carbon instability.

continued

> **BOX 5** *continued*
>
> While the session focused on larger scales, speakers also emphasized the importance of landscape-scale models that can test the efficacy of different management decisions, while taking into account future climate and CO_2 scenarios, in order to support policy and decision making. Breakout discussions also pointed to the importance of scaling up landscape- and regional-scale models, which may benefit from additional computational resources. Several participants noted the current disconnect between model parameters at different scales—for example, vegetation, fuels, aerosols—and the importance of focusing on local and regional modeling to improve assessments, rather than only focusing on the global scale. Another breakout suggested representing the heterogeneity of land cover in global models with a statistic or metric.

Future Management to Support Net-Zero Targets

The third and final day of the workshop focused on accounting for wildfire emissions as part of country-level greenhouse gas (GHG) inventories and the range of mitigation options to reduce future wildfire emissions and increase resilience of ecosystems to climate change impacts.

ACCOUNTING FOR WILDFIRE EMISSIONS IN NATIONAL REPORTING AND NET-ZERO TARGETS

Werner Kurz, Natural Resources Canada, motivated the discussion toward achieving net-zero targets by noting that to limit warming to 1.5°C, net-zero anthropogenic (human-driven) emissions must be reached by 2050, and emissions must be net negative during the second half of the century. Achieving net-negative emissions means that anthropogenic carbon removals would be greater than anthropogenic emissions, and there is an expectation that the land sector—particularly forests and wood product carbon storage—will contribute to these removals; however, these forests are at risk from climate change impacts, including from fires.

In the first session, speakers introduced the impacts of wildfire emissions on national GHG reporting and the implications for net-zero targets through three country-specific case studies. Speakers described the magnitude of GHG emissions from wildfires in Canada, the United States, and Australia and the ways these emissions are accounted for and reported.

National Emissions and Reporting

Parties to the United Nations Framework Convention on Climate Change (UNFCCC)—198 countries—self-report their national GHG emissions; these reports can then be used to evaluate whether countries are meeting their emission reduction targets. However, national accounting of net GHG emissions only includes direct anthropogenic emissions and removals, Kurz explained. Under the Intergovernmental Panel on Climate Change (IPCC) reporting guidelines, countries can separate emissions from forests into "managed" or "unmanaged" categories.[1] In Canada—the focus of Kurz's remarks—there are 226 million hectares of forest for which emissions and removals have been reported annually since 1990, but approximately 121 million hectares of unmanaged forest where only land

---

[1] The IPCC defines managed land as "land where human interventions and practices have been applied to perform production, ecological or social functions" (IPCC, 2006). IPCC guidance provides latitude for governments to refine the definition of managed land to meet their national circumstances (Ogle et al., 2018).

use changes are monitored and reported. In Canada, there is large interannual variability in emissions from managed forests driven by area burned, and there has been an increase in GHG emissions from managed forests such that they have transitioned from a carbon sink to a source over the past two decades (Figure 16).

The interannual variability in Canada is driven by natural processes ("disturbances")—which are being influenced by climate change—while anthropogenic emissions from direct human influences have steadily increased over time. The question from an emissions and management perspective is whether mitigation can bend the anthropogenic emissions curve or if changes in natural processes due to climate change will overwhelm any mitigation actions (Figure 16). Prior to 2023, 2021 had been the worst fire season in Canada since 1990 with direct fire emissions of 290 million tons of carbon dioxide equivalent (CO_2e). In 2023, as of the time of the workshop in mid-September, the fire emissions were roughly three times larger than emissions in 2021, and emissions from managed forests alone were 1.4 times the GHG emissions from all other sectors in Canada combined.

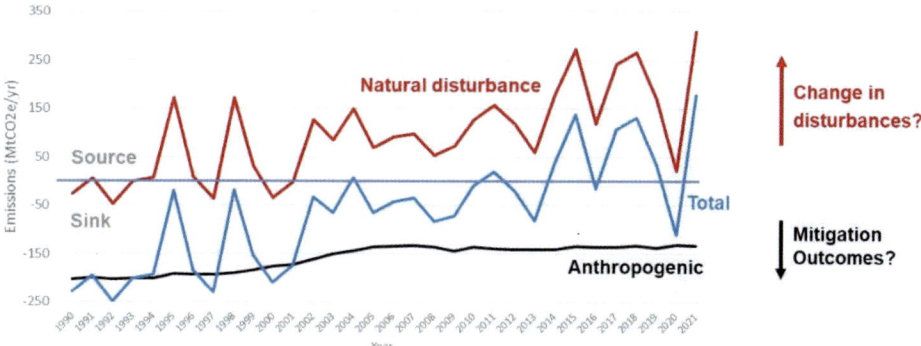

FIGURE 16 Wildfire greenhouse gas (GHG) emissions and removals for managed forests in Canada from 1990 to 2021. Negative emission values mean the forests are a carbon sink, and positive emission values mean forests are a carbon source. Emissions are disaggregated into those from managed forests predominantly affected by direct human (anthropogenic) activities (black line) and by natural processes (disturbances), influenced by climate change (red line). The blue line shows the total, or net (emissions plus removals), GHG emissions. Emissions are shown in million tons of carbon dioxide equivalent ($MtCO_2e$). SOURCE: Adapted from Environment and Climate Change Canada (2023).

The recent fires in Canada demonstrate the differences between mitigation actions, which affect a small proportion of the forest in some years, and impacts from climate change, which affect all forests every year. Kurz explained that analyses of the last 20 years have shown that disturbance impacts consistently exceed mitigation benefits by one to two orders of magnitude or more. In addition to the direct emissions from fires, the decay of fire-killed trees also contributes to indirect GHG emissions in the years after fires.

Grant Domke, U.S. Forest Service, focused on U.S. wildfire emissions from the land sector from 1990 to 2021. In the United States, the land sector[2] is a sink of carbon and offsets the equivalent of approximately 13 percent of economy-wide emissions. The strength of this carbon sink, primarily from the forest land sector, has declined in recent years in part due to the frequency and magnitude of wildfire emissions (EPA, 2023). In addition to emissions of carbon dioxide (CO_2) wildfires are the largest source of nitrous oxide (N_2O) and second-largest source of methane in the land sector (EPA, 2023).

For the purposes of reporting GHG emissions to the UNFCCC, of the land base in the United States, about 95 percent is considered managed—essentially the entire contiguous United States (CONUS) and most of Hawai`i, with forests representing one-third of the managed land; all of the unmanaged land is in Alaska (Ogle et al., 2018). Like Canada, the United States only reports emissions associated with managed lands, though they do compile (but do not report) emissions and removals for the unmanaged lands in Alaska. U.S. wildfire emissions are compiled using the Wildland Fire Emissions Inventory System, which uses burned area from the Monitoring Trend and Burn Severity records, observations from the MODIS (Moderate Resolution Imaging Spectroradiometer) satellite, and an interagency dataset together with other modules to estimate combustion and emissions per fire. Figure 17 shows annual emissions and area burned from 1990 to 2021 in which there is large interannual variability, like for Canada, including the contribution of fires on managed land in CONUS versus Alaska. Domke also highlighted that quantifying prescribed fire on managed lands in the United States remains a large uncertainty, and noted that better integrating and harmonizing different emission estimates would be important, particularly as prescribed fire is considered as a management tool.

David Bowman, University of Tasmania, discussed carbon accounting in Australia where there have also been massive emissions from fires recently; in particular, the 2019–2020 fires were estimated to be 1.6 times Australia's national anthropogenic emissions for 2019 (Crippa et al., 2020). However, these large wildfire emissions were not captured as part of Australia's national "net" emissions. Australia reports their emissions following the UNFCCC guidelines for the land use, land use change, and forestry (LULUCF) sectors in which GHG emission calculations vary by vegetation type and whether the land is managed or unmanaged. For example, deserts are considered in the guidelines to be unproductive, so even though they do burn in Australia, those emissions are not counted.

National GHG inventories are designed to reflect human-induced emissions. A number of assumptions are built into the UNFCCC guidelines: fire disturbances are assumed to have a transient effect on GHG emissions; carbon stocks are assumed to rapidly recover and return to pre-fire levels; fires on managed land are assumed to be patchier, less severe, and different from unmanaged wildfires; and historically and geographically, anomalous wildfires in forests have been excluded from national anthropogenic emissions estimates because they are assumed to be beyond human control (Bowman et al., 2023). Major unmanaged landscape fires are disaggregated from Australia's national total emissions and reported separately under the "national disturbance provision."

---

[2] The land sector includes emissions and sinks related to land use, land use change, and forestry.

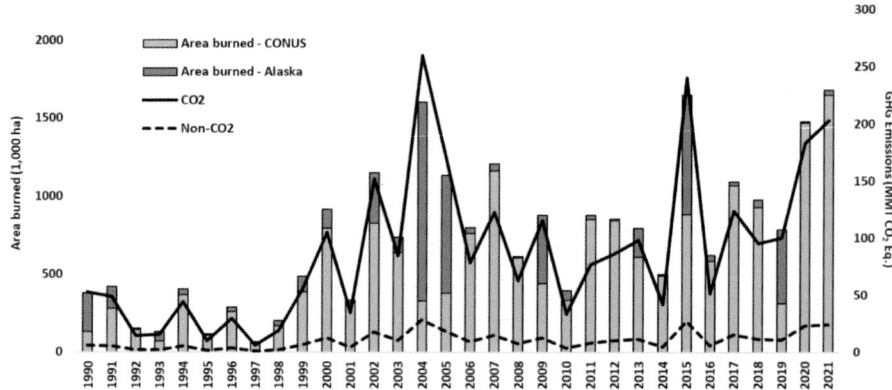

FIGURE 17 Estimated area burned (bars) and greenhouse gas (GHG) emissions (lines) from wildfire and prescribed fire on managed forest land in the contiguous 48 states (CONUS) and Alaska from 1990 to 2021 reported to the United Nations Framework Convention on Climate Change. The solid line shows carbon dioxide (CO_2) emissions, and the dotted line shows emissions from nitrous oxide (N_2O) and methane (non-CO_2). NOTES: ha = hectare; MMT CO_2 Eq. = million metric tons of CO_2 equivalent. SOURCE: Domke et al. (2023).

Opportunities to Improve Emissions Accounting Moving Forward

Speakers explained that moving forward, it would be important to quantify emissions from natural processes that are not currently a part of national GHG inventory reporting. Domke echoed the importance of accounting for emissions and removals on both managed and unmanaged lands. Kurz explained that in Canada, the largest limiting factor for including unmanaged land in emission inventories is the limited forest inventory data, though Canadian scientists are making advances using remotely sensed data. The United States does estimate unmanaged land in Alaska, and Domke explained that they are using data on managed land to inventory unmanaged land in the interior of Alaska. Bowman added that GHG emissions accounting can be used to incentivize forest management, as has been done in tropical savannas in Australia. However, not accounting for major unmanaged fires as part of national reporting disincentivizes management efforts to avoid these types of fires.

The quantification and reporting of these emissions raise the question: who is responsible for the direct and indirect GHG emissions from wildfires driven by climate change? Kurz explained that because nonanthropogenic emissions from fires are not reported and accounted for in national net-zero emission targets, global emissions would have to be further reduced in other sectors to truly achieve net-zero GHG emissions. For global forests to contribute to net-zero targets, Kurz argued for shifting the focus from timber management to carbon management with the goal of increasing carbon sinks and forest resilience to future climate change impacts. Domke called for holistically considering all lands and communities when moving toward the idea of carbon stewardship on the landscape.

Kurz noted that there is an expectation based as part of many countries' nationally determined contributions that the forest sector will contribute toward national net-zero targets. However, the effects from climate change on forests in many countries are such that emissions are overwhelming the ability to manage and maintain forests to enhance carbon sinks. Domke reiterated that in both the United States and Canada, forests have historically been a strong carbon sink; however, rather than assuming that forests will maximize their capacity as a carbon sink in the future, a more realistic goal may be to maintain current forest sink capacity in some areas. Kurz suggested changing the mental model of forests as a mechanism to store carbon in the landscape—because of the potential release from wildfires—to a focus on forests as a mechanism to remove CO_2 from the atmosphere.

Regarding management options for forests in Canada in the context of intensifying fire regimes, Kurz highlighted that there is no single silver bullet and emphasized the challenges of scale and large differences in forests across Canada with diverse histories of fire and fuel management. Kurz offered three main management strategies that could make a difference: reducing fuel loads, reducing flammability (i.e., switching from coniferous species of higher timber value to broadleaf species of lower timber value and reduced flammability), and changing the horizontal structure of the landscape (i.e., creating more open spaces with less risk for continuous crown fire). Kurz also recognized the important role of increasing prescribed and cultural burning.

OPPORTUNITIES TO REDUCE FUTURE WILDFIRE EMISSIONS IN DIFFERENT BIOMES

The remainder of the workshop was centered around the solution space for reducing GHG emissions from wildland fires. Speakers in the next session focused on different biomes to provide specific examples of management opportunities: western United States temperate forest, North American boreal forests, tropical peatlands, tropical forests, and the range of ecosystems across Australia. Breakout discussions likewise considered climate-effective, socially inclusive, and ecologically appropriate mitigation efforts to reduce future wildfire emissions in Arctic/boreal, temperate, and tropical biomes. A clear theme throughout the session was the importance of balancing the desire to reduce emissions with cultural practices and norms and the importance of any mitigation effort to include co-planning and co-management with local Indigenous communities (Box 6).

Western U.S. Forests

Paul Hessburg, U.S. Forest Service, Pacific Northwest Research Station, focused on solutions in the western U.S. interior forest landscapes that were historically and continue to be sculpted by fire. Echoing context outlined by previous speakers, Hessburg highlighted the tens of millennia history of lightning and Indigenous ignitions in the region that created large areas of open forest, wet and dry meadows, and sparse woodlands and shrublands (Lake and Christianson, 2019). Indigenous burning focused on closed canopy forests to make food and resources available (Maclean et al., 2023; Roos et al., 2022; Swetnam et al., 2016), but absent these fires, forests have grown denser, and many previously unforested areas are now forested (Hagmann et al., 2021). Indigenous and lightning fires burned in moist and cold forests and often as much as 35–50 percent of a large landscape area was burned or recovering after fires (Hessburg et al., 2016, 2019). The continental United States is now in a large fire deficit in terms of area burned compared to historical pre-industrial burning (Abatzoglou et al., 2021; Leenhouts, 1998; Parks et al., 2015).

BOX 6
Co-management of Wildfires and Emissions with Indigenous Communities

Session speakers emphasized that working with and respecting Indigenous and cultural practices of local communities is a critical part of any management strategy. Paul Hessburg pointed to the importance of co-planning and co-management with Indigenous communities to restabilize forest conditions in the western United States and noted that dialogue is critical because fire management practices often vary by tribe, band, or family group. Geoff Cary added that because of large differences in landscapes and fire regimes in Australia in some places, the long-term reintroduction of frequent cultural burning may be advantageous with respect to reducing GHG emissions (Figure B), whereas other places may need a long-term absence of fire. Because the flammability of landscapes—for example, in the Xingu Basin in the southeastern Amazon—has changed so much, Marcia Macedo pointed to the need for co-learning about how to use fire for both cultural and fire control purposes to achieve shared goals. Peter Frumhoff added that new projects aiming to co-design management strategies can serve as testbeds and models for how to co-design and co-produce knowledge for fire management.

FIGURE B Indigenous fire mosaic in the western desert of Australia. After being removed from their land, the Martu people reintroduced fire after a 15-year time period, bringing back a landscape mosaic. Martu burning achieved nearly all seral states (early to late). SOURCE: Don Hankins.

The historic regime of frequent low- or moderate-severity fire was an important stabilizing feedback that led to forest conditions that were in sync with the climate and native vegetation and improved the likelihood that the next fire would also be low to moderate severity (Figure 18). Forest reburning is an essential large landscape-stabilizing feedback (Hessburg et al., 2019; Povak et al., 2023; Prichard et al., 2017). In the historical regime, fires of varied sizes and severity—that overlapped in space and time—created shifting mosaics of forest and nonforest conditions that self-regulated by dampening fire sizes and their severity. Additionally, nonforest areas within large landscapes (25–75 percent of large landscapes as shrubland, sparse woodland, wet and dry meadows, wetlands) also limited future fire size and severity.

Without these high-frequency fires, trees quickly accumulate, allowing flames to climb the layered subcanopy, resulting in the crown fires experienced in the western United States today (Figure 18). Hessburg explained that the shifting reburn and recovery mosaic of the past is the missing ingredient today, and the current forest cover in the interior West is not sustainable for forested ecosystems. Current conditions have made the interior West more vulnerable to fire: increased canopy fuels intensify energy available for severe fires; increased connectivity of fuels creates opportunities for large and spreading fires; and changes in climate and weather increase fuel curing and area burned, meaning high burn severity conditions are readily available at regional to provincial scales.

FIGURE 18 Illustrations of historical and current fire regimes in western U.S. temperate dry and moist mixed-conifer forests. (Left) The historical fire regime included frequent low- and moderate-severity fire (top) that led to continuously stabilized, open-canopy forest conditions (bottom) and the increased likelihood of future low- and moderate-severity fires. (Right) In the current low-frequency fire regime driven by fire exclusion, forests accumulate dead wood, fuel ladders, and trees, and fires now readily reach the crowns of overstory trees thereby creating large areas of fire-killed forest.
SOURCE: Robert Van Pelt.

The change agents beginning in the 1800s up to the present day included fire exclusion (e.g., reduced Indigenous burning, land development for agriculture and urban settings, fire suppression), timber harvest, climate change (e.g., increased temperatures and wind, longer fire seasons, reduced snowpack), and smoke management (e.g., regulations to reduce intentional burning). With this context, Hessburg highlighted a number of potential solutions to reduce future wildfire emissions in interior western U.S. forests:

- Increasing open-canopy forest conditions (i.e., less fuel accumulation) (Figure 18) and reestablishing burned and recovering mosaics of forest and nonforest conditions;
- Stabilizing the competing factors that grow and remove forests through management that would, for example, restore nonforests, hardwood forests, and wetlands, and open-canopy forest conditions; and
- Restoring the positive ecological role of fire by incorporating Indigenous knowledge, practice, and management leadership, using the tools of cultural and prescribed burning, forest thinning, and biomass removal that can contribute to a bioeconomy.

Breakout discussions focused on management in temperate forests broadly. One breakout echoed the importance of increasing prescribed and cultural fires and pointed to a strategy of strategically targeting ecosystems based on either the risk of carbon loss or the vulnerability of the ecosystem itself to loss from fire. Relatedly, other participants discussed the conditions that could be helpful to transition to an adaptive management fire regime, including increased public awareness and comfort with prescribed burning (and the resulting smoke) and shifting frameworks for forest management to choosing when to have fire and smoke, not whether to have it. Discussions also recognized conflicting land management priorities and noted the importance of supporting and incorporating Indigenous practices and cultural burning and potential opportunities to incentivize land managers to accomplish mitigation goals. Participants recognized the heterogeneity of temperate landscapes that involve different approaches and the importance of examining local examples of adaptations to the social and environmental conditions of a place, for example, from Indigenous communities.

North American Boreal Forests

Peter Frumhoff, Harvard University and Woodwell Climate Research Center, focused on fires in the Alaskan and Canadian boreal forests. The burned area of wildfires in Alaska and Canada have roughly doubled since the 1960s (Figure 19) (Phillips et al., 2022). This increase is strongly linked to climate change, and these trends are expected to continue without effective fire management interventions at a large scale. Phillips et al. (2022) estimated that wildfires in Alaska and Canada could contribute to a cumulative net source of up to 12 gigatons of net CO_2 emissions by 2050 (Figure 19), which may be a conservative estimate in light of Canada's 2023 fire season (on the order of 2 gigatons CO_2 equivalent at the time of the workshop).

Wildfires in Alaska and other boreal forests are not currently managed with the goal of limiting carbon emissions. Current fire management practices in Alaska focus on protecting lives and property and are geographically prioritized by fire management zones in which the greatest suppression efforts go toward fires in the "critical" zone. Phillips et al. (2022) found that the fire management zone explains approximately 22 percent of the total variability in fire size in Alaska. When management data on effectiveness in reducing emissions are tied to economic data for the cost of fire suppression, fire management in Alaska appears to be a viable and cost-effective approach to keep carbon in the ground and out of the atmosphere. On the basis of this analysis, Frumhoff proposed that one approach to limiting GHG emissions in the North American boreal forests may be to target fire suppression for carbon protection early in the season before fires become large in order to support the overarching goal of maintaining wildfires at historically or ecologically sustainable, pre-climate change levels.

This management approach is being piloted in the Yukon Flats National Wildlife Refuge in Alaska by the Bureau of Land Management and the Fish and Wildlife Service in which they are designating a change in the fire management regime for 1.6 million acres across the refuge that are underlain with carbon-rich Yedoma permafrost. The goal of the project is to slow the release of GHGs to the atmosphere and reduce air pollution while maintaining wildfire habitat diversity. Frumhoff noted that this field experiment will help determine the potential for targeted suppression of boreal wildfires to limit GHG emissions.

To the question of managing boreal wildfires more broadly, Frumhoff pointed to the other impacts of wildfires (e.g., black carbon deposition on snow and ice, smoke inhalation and consequences for human health), and recognized the opportunity to consider other co-benefits—beyond reducing GHG emissions—of wildfire management that may motivate different interested and affected groups and policymakers to limit wildfires (e.g., co-benefits for aviation, tourism, infrastructure). Frumhoff also made the point that fire management in the boreal will require real investments. According to one estimate, managing Alaska wildfires to pre-climate change levels would require an average annual investment of $700 million through 2030, roughly five times the current level of annual investment in Alaska wildfire management (Phillips et al., 2022).

Frumhoff concluded by making the point that expanded wildfire management to limit GHG emissions in the boreal only makes sense in the context of broader national and international commitments to deep decarbonization. While limiting wildfire emissions and maintaining carbon in the landscape would help buy time, warming and wildfire intensity ultimately will continue to increase until global emissions reach net zero and net negative. For this reason, Frumhoff urged caution around using carbon markets as the primary source of financing for limiting boreal wildfires so as not to inadvertently undermine the importance of deep fossil fuel emission reductions. Frumhoff also underscored that any consideration of altered fire management in the boreal forests would need to fully engage and consider the perspectives and priorities of Indigenous communities who have long been the stewards of the landscape but have too often been left out of the design of fire management plans and priorities.

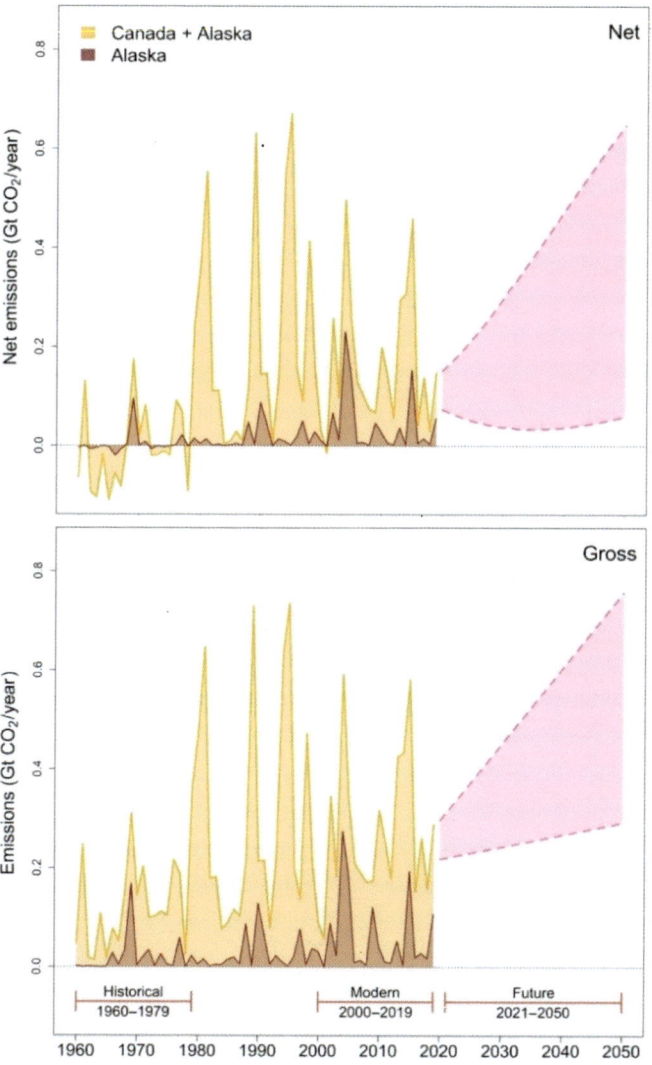

FIGURE 19 Observed (1960–2019) and projected (2020–2050) gross and net (accounting for forest regrowth) carbon dioxide (CO_2) emissions from boreal Alaska (brown) and North America (both Alaska and Canada) (orange). Pink areas show the range of projected average annual emissions for boreal Alaska and Canada through 2050. Projections were derived from the literature, and the range shown represents the upper and lower quartiles of these projections. SOURCE: Phillips et al. (2022).

Breakout discussions that focused on management in Arctic and boreal regions highlighted additional challenges and opportunities in the region. Discussions noted the unique challenges of managing fires in the Arctic/boreal region, including the large scale and remoteness, limited resources and infrastructure, large interannual variability, geopolitical challenges (i.e., Russia representing two-thirds of the Arctic/boreal zone), and the rapid pace of warming. While there are a variety of different management options—cultural burning, fuel and vegetation treatment, prescribed burning, forest regeneration, wetland protection—the challenge is understanding and agreeing on when and where to apply these strategies in consultation with communities. Other breakout discussions recognized the importance of education and outreach to the public and policymakers and pointed to the possibility for economic incentives to stop inhibiting deciduous growths. Finally, many participants noted that management solutions likely involve broader policies that both reduce global fossil fuel emissions and consider carbon and climate as part of fire management.

Tropical Peatlands

Susan Page, University of Leicester, focused on peatland fires in the tropics that are highly vulnerable and fire-prone systems today. The resilience of tropical peatland is primarily determined by hydrology where a high water table maintains moist peat and aboveground biomass such that fire is not a natural feature of these ecosystems. Prior to development in Southeast Asia and other intact tropical peatlands, there was no or low fire risk. Recently, the additions of agriculture and people—for example, people from other islands moving into Indigenous peoples' homeland in Indonesia—have lowered the water table, exposed a large fuel base, and introduced a range of ignition sources, changing the fire regime of tropical peat systems. Additional drivers that have made peatlands more fire prone include increased drainage and incidence of drought.

Fire is used as a traditional land management tool across Indonesia, though use of fire at scale for brush clearing and plantation development has increased over the last two to three decades due to migration of people into these areas from other places. These trends have coincided with forest degradation, forest loss, and peatland drainage that together have made the landscape increasingly flammable. The government of Indonesia has implemented fire bans, controls, and regulations on peatland drainage. However, the use of fire on these flammable, remote landscapes can result in out-of-control fires that are challenging to extinguish. The smoke emissions from fires in Indonesia can impact people as far as Thailand and the Philippines, and these large fires have a large impact on the economy and people's health in Indonesia (e.g., Koplitz et al., 2016). To manage peatland landscapes, Page offered some potential solutions:

- Reducing fire risk by focusing on the use of fire only when necessary and encouraging low or no fire land management (e.g., Fire Free Village initiatives that financially incentivize low or no fire);

- Re-wetting the landscapes (deliberate actions that aim to restore the water table of a drained peatland) while also maintaining livelihood security for communities that depend on growing dryland crops;
- Using fire danger warning systems to limit the risk of accidental ignition;
- Implementing active and ongoing fire management for landscapes where there is forest regeneration and replanting; and
- Maintaining wet peatlands by finding alternative pathways for economic sustainability that do not involve development that disrupts hydrology in peatlands.

Tropical Forests

Marcia Macedo, Woodwell Climate Research Center, emphasized the importance of tropical forests as stores of aboveground biomass carbon. Recent estimates suggest that removing global forests would raise global temperatures by as much as 1°C, and the Amazon—the focus of Macedo's work—represents about 50 percent of global forest area. The Amazon is in a state of stress; climate change and fire will be determining factors in whether tropical forests become more resilient or more degraded (Castanho et al., 2020; Trumbore et al., 2015). As in tropical peatlands, fire is not a natural part of the ecosystem in tropical forests; the main ignition sources for these fires are anthropogenic. The primary drivers of fire activity in the Amazon are fuel, ignition, and climate, and all three drivers are currently changing (Alencar et al., 2020).

The priority solution from Macedo's perspective is stopping deforestation in the Amazon because it would address all the drivers of fire activity. Deforestation adds a new fuel source (i.e., accumulated dead biomass) to the landscape that has a high probability of ignition, emits GHGs, and changes local and regional climate. Deforested areas are roughly 5°C warmer than forested areas and recycle 30 percent less water to the atmosphere, which enhance conditions for fire. Managing forest degradation is an important strategy for reducing emissions from tropical forests (Kruid et al., 2021). The challenge is attributing the degradation to different drivers to identify intervention strategies (Figure 20). Data and modeling can also be tools for fire management in the Amazon. For example, forecasts ahead of and during active fire season can help identify locations where efforts are needed to stop fire from escaping into wildlands. Near-real-time data about fire type, for example, from the Amazon Fire Dashboard,[3] can provide managers with critical information.

To emphasize the importance of wildland–agriculture interfaces, Macedo shared the example of the Xingu Basin in the southeastern Amazon where land management fires occur adjacent to large blocks of forest. Land management fires can be an ignition source, so reducing unnecessary fire use for pasture and agricultural management could help protect forested areas. Understanding the interactions between the compounding disturbances of fire and drought in this region is another challenge and piece of the solution space (e.g., Balch et al., 2015). In the Xingu Indigenous territory, every time there is a major drought, 10–15 percent of the area burns. While some burning is prescribed, roughly 15 percent of

---

[3] See https://amzfire.servirglobal.net.

the area is now a grass-dominated system that is fire prone and perpetuates the cycle of fire (Silvério et al., 2022).

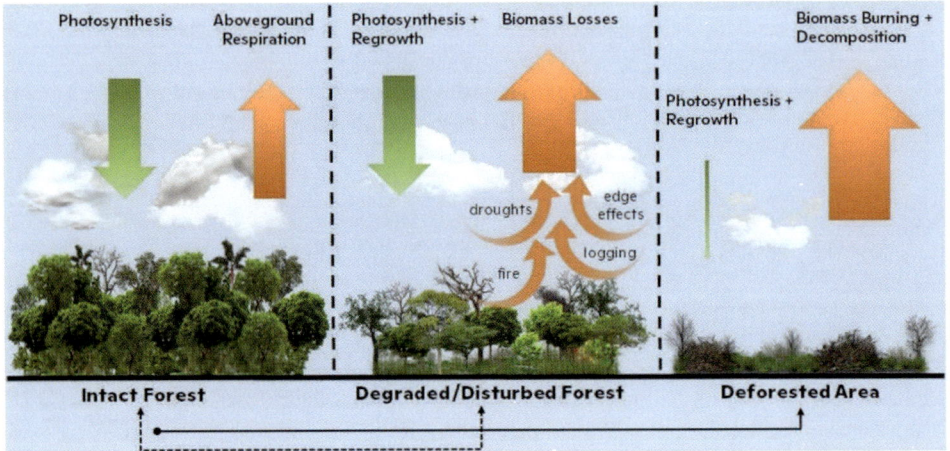

FIGURE 20 Conceptual representation of aboveground carbon fluxes typical of healthy forests (left), degraded or disturbed forests (middle), and newly deforested areas (right) in the Amazon. Green arrows show gains in carbon stock and orange arrows show losses. The middle panel shows indirect and direct anthropogenic drivers of degradation and disturbance. The one-way solid black line shows deforestation and conversion into another land cover that is unlikely to regenerate. The two-way dashed arrow shows the degradation/disturbance of intact forests that may regenerate depending on the intensity and frequency of disturbances. SOURCE: Kruid et al. (2021).

Macedo also underscored that integrating traditional knowledge about fire use and impacts is crucial. As an example, researchers worked with communities in the Xingu reserve to study 60 useful species (e.g., used for construction or medicinal purposes, sold for reforestation) and found that about one-half of the species fruit during the late dry season when prescribed burns typically happen. Disruptions to harvesting these fruit species from prescribed burns was an unintended consequence of the management practice. This finding was shared with the fire management agency and is an example of co-developing research questions and adapting fire management strategies to benefit local communities.

Finally, Macedo highlighted additional levers that can be used to reduce fire on the landscape. First, the Amazon has large tracts of undesignated public forest that have poor governance and are prone to land speculation and fire use for deforestation, and so clarifying land tenure could reduce fire. Second, streamlining data flows and building capacity and data tools could support firefighters. Finally, supporting Indigenous land rights and leadership in the Amazon is critical, Macedo said.

Breakout discussions considered management strategies across different tropical ecosystems. Because land use change is the primary driver of fire in tropical systems, any intervention to reduce wildfire emissions likely involves working with people and local communities. Participants discussed the goal of a participatory process with shared leadership to permit the use of fire in traditional and smallholder systems while minimizing the risk of escaped fire in humid tropical forests. Beyond community fire management, participants also discussed the inclusion of local people in academic research teams, industries, and legislative and policy discussions around fire management as well as the need for long-term investments, including for ongoing monitoring. Another unique characteristic of tropical ecosystem management is that reducing fire emissions would also mean altering human activities and thus reducing financial benefits. Consequently, many participants noted the opportunity to associate tropical forests and peatlands that should not be burned from an ecosystem health and emissions standpoint with benefits or compensation. Additional barriers to implementing management policies at scale in tropical systems include political boundaries, sovereignty, instability of governance, and corruption.

Regarding potential future research areas, participants discussed opportunities to anticipate changes in risk by tracking and forecasting seasonal and interannual changes in fire risk that reconnect tropical landscapes, permitting large, destructive fires in remote regions. Tropical ecosystems could also benefit from improved real-time monitoring of low-intensity fires, including smoldering fires in peatlands or escaped fires in tropical forests, using remote sensing tools and ground-based measurements. Several participants highlighted the opportunity to combine a series of monitoring systems that could allow for local reporting of escaped fires without risk of penalty to encourage participation and collaboration with local communities.

Australia

Geoff Cary, Australian National University, spoke about management of wildfires across the diverse landscapes of Australia. As many of the speakers discussed, wildfire management to reduce GHG emissions likely involves changing the fire regime to one that has intrinsically lower emissions; protects carbon, trees, and carbon-sequestering forests; or both. Cary shared the example of management of northern tropical savannas in Australia in which prescribed, lower-intensity burning early in the annual dry season can offset the area burned by higher-intensity, more extensive fires later in the dry season and significantly reduce non-CO_2 GHG emissions (e.g., Edwards et al., 2021; Russell-Smith et al., 2013).

Management of temperate eucalyptus forests is also critical, Cary explained. For example, during the major fires of 2019–2020, approximately 18 million acres of temperate eucalyptus forests and woodlands burned in eastern Australia. Unlike in tropical savannas, prescribed burning is a more limited management solution for reducing GHG emissions (i.e., around 4 acres of prescribed burning is required to offset 1 acre of high-intensity wildfire over the next 5 years) (Bradstock et al., 2012). An alternative way to reduce wildfire extent, as suggested by Cary, is to reduce unplanned ignitions, by either preventing or rapidly suppressing ignitions when they occur, though significant advances in ignition detection and suppression technology would likely be required. Cary shared results from a modeling study that examined the relative importance of various controls on fire in temperate or similar landscapes and found that interannual variability in weather was most important for predicting variations in area burned, followed by reducing or suppressing ignitions (reduced wildfire area in Australia model systems by >80 percent), and finally, prescribed burning (reduced wildfire extent by 25–40 percent).

Cary also considered management options to protect the carbon storage and sequestration potential of forests. For example, eucalyptus trees can store large amounts of carbon, and while most species can survive and recover from fire, some species can be killed by fires intense enough to scorch the canopy and require a fire-free period of over 10 years to regenerate (Gale and Cary, 2021). Cary discussed applying prescribed burning at the edge of these regenerating eucalyptus forests as a potential strategy for protecting carbon-sequestering eucalyptus forests.

THE SOLUTION SPACE AND EXAMPLES OF NEXT STEPS: FOREST MANAGEMENT OF TOMORROW AND LIVABLE EMISSIONS

The workshop concluded with a moderated conversation among practitioners about the solution space and examples of potential next steps for ecosystem and fire management in light of the opportunities and challenges for reducing GHG emissions identified throughout the workshop. The discussion centered around several themes that brought together threads from throughout the workshop: working with local communities; co-benefits and trade-offs between carbon, ecosystems, and livelihoods; tools for management; and the management horizon for the future.

Working with Local Communities

Jimmy Fox, U.S. Fish and Wildfire Service, manages the Yukon Flats National Wildfire Refuge in interior Alaska, which overlays the traditional homeland of the Gwich'in Athabascan people. Today, about 1,200 predominantly Indigenous people live on private lands within the boundaries of the 8.6-million-acre refuge. The Yukon Flats has 6,000 years of fire history, but the frequency of fire has increased in the past several decades, and most fires on the refuge are monitored but allowed to burn. However, these fires can threaten human health, subsistence lifestyles, natural diversity, and global climate goals. Fox elaborated on the pilot project, introduced previously by Frumhoff, in which the U.S. Fish and Wildlife Service and the Bureau of Land Management's Alaska Fire Service have agreed to suppress wildland fires through the first half of the fire season over 1.8 million acres, roughly 20 percent of the refuge. The goals of the project are to preserve mid- to late-successional plant communities for habitat diversity, protect the estimated 1.1 Gt of carbon in Yedoma permafrost that underlies the refuge, and reduce air pollution and its impacts for the residents of the refuge.

Fox said the key to working with local communities is to show up, listen, and engage thoughtfully. Instead of a single strategy for engagement, Fox highlighted the importance of developing location- and tribal-specific approaches, which require investments of time and resources over a sustained period of time. One lesson Fox has learned is to look for ways to engage that work for local communities rather than prioritizing his own interests as a federal land manager and to be flexible and responsive to the ways in which tribes want consultation to look.

Co-benefits and Trade-offs between Carbon, Ecosystems, and Livelihoods

Paul Hessburg, U.S. Forest Service, shared examples of co-benefits from reestablishing landscape resilience by reducing fuel loads and reintroducing the ecological role of fire in western U.S. temperate systems. Reestablishing forest resilience could stabilize CO_2 uptake and storage in forests, albeit at lower levels, and protect municipal water supplies and other vital infrastructure. The wildland fuel restoration problem and the community resilience problems are quite independent and involve different mindsets and tactics (Calkin et al., 2023). Reducing wildfire vulnerability overall could improve human health (from less smoke) and protect human infrastructure. Hessburg suggested that inviting fire back into what are inherently fire ecosystems and moving beyond the fire suppression era would restore resilience to a future climate and shifting wildfire regimes.

Dan Thompson, Canadian Forest Service, provided an example of weighing the trade-offs between animals, livelihoods, and fire management strategies in Canadian boreal forests. In Canada, woodland caribou are a federally listed species at risk, and one of their primary habitats is open, old conifer forests, which are also the most flammable fuel type. Boreal forest conversion from conifers to broadleaf may be appropriate in some locations, such as near communities, but Thompson explained that there is a trade-off in habitat spaces for species at risk. Adding complexity to the management space are communities, including Indigenous communities, who run traplines for fur-bearing mammals in old conifer forests, meaning any forest conversion could also impact their livelihoods. Thompson suggested that instead of forests that are uniformly broadleaf or uniformly old conifer with fire suppression, a balanced solution may look like a mosaic of landscapes and management approaches, drawing from the traditional management model in the past Canadian boreal. A mosaic in the boreal may look different from a mosaic in temperate forests. For example, a mosaic could include pockets of old conifer forests in places that are advantageous to wildfire and away from people who would be impacted by smoke, areas broken up with lakes and islands, and the introduction of broadleaf in other areas.

Natasha Ribeiro, University of Eduardo Mondlane College of Agriculture and Forestry, spoke about the miombo woodlands in Mozambique, one of the largest dry forest ecosystems in the world. Fire has historically been part of the miombo ecosystem for more than 200,000 years. The Niassa Special Reserve, an area covering 42,000 square kilometers in northern Mozambique, has a typical fire frequency of 3–4 years; however, climate change and population growth have resulted in an annual fire frequency in some places (45 percent of the reserve), which is changing forest composition and structure. About 65,000 people live within the Niassa Special Reserve and use fire and other forest resources to maintain their livelihoods. The ecology of the miombo is determined by people, fires, vegetation, and animals, and is a challenging ecosystem to manage because of its size and diversity of landscapes. Ribeiro called for an integrated fire management policy that would include traditional knowledge and the importance of fires in the miombo, instead of a zero-fire policy that ignores traditional knowledge and practices.

Tools for Management

Thompson said that in Canada, while there are some gaps in data, models, and observational tools needed to ideally implement and monitor solutions, they are close to having the needed tools. Even though Canada is not as densely observed as parts of Europe or the continental United States, particularly in the north, the forest is low diversity and relatively well characterized for the purposes of management. Improving prediction tools—for example, fire extent and fire growth 3–4 days into the future—are marginal improvements, and Canada is not in a place of information scarcity. Instead, Thompson explained that the real challenge for management is the extent to which climate change has disrupted the equilibrium of the boreal fire ecology. The boreal forests that are relatively old (~100-year median age) are now burning at a rate and frequency that is out of equilibrium with its natural state due to amplified warming at higher latitudes.

Thompson identified the need to provide managers with actionable information, rather than a flood of data that may not be practically useful. Thompson also noted that real-time information needs for the public may be changing. For example, many eastern Canadian community institutions (e.g., schools, daycares, outdoor sports leagues) outside of major cities do not have protocols in place to deal with severe air quality advisories as issued by the federal government, as wildfire smoke and air quality more broadly have not traditionally been an issue.

Jayaprakash Murulitharan, Cambridge University and Ministry of Environment, Malaysia, shared his perspective both as an air quality researcher and from his experience in government in Malaysia and with ASEAN (Association of Southeast Asian Nations). Murulitharan centered on the historic event in 2015 when Southeast Asia experienced severe air pollution (or haze) due to wildfires. Malaysia is affected from wildfire smoke from both the north (from agricultural burning in Thailand, Laos, and Burma) and the southwest (from peatland burning in Sumatra and Kalimantan). Over the past 7–10 years, there has been a focus on real-time and efficient data sharing between countries in Southeast Asia and efforts to establish a coordination center to ensure that fires are reported early and that weather predictions can be used to help countries prepare.

Murulitharan also noted that since 2015, there were improvements to peatland management in Indonesia, but this required a strong political commitment from the top down. Moving forward, Murulitharan said that there is a need to highlight the scale of wildfires in the region and their impact on climate to move toward wildfire mitigation and peatland management throughout the region.

However, countries in Southeast Asia have not typically made the connection between climate change and haze, Murulitharan explained. One challenge is that when fires are periodic events, there can be a focus on blaming individual countries where fires originate; a second challenge is that the focus has been on managing fires and haze in real time, rather than translating the impact of fires on GHG emissions or the role of climate change. There is not currently harmonized data sharing between countries because each individual country has their own approach to gathering data and quantifying emissions. While there have been advances in making connections between smoke and health with the public, the translation of the impact of fires on climate change is missing.

Management Horizon into the Future

Speakers concluded the session discussing examples of goals for management and policies in the future. Hessburg explained that inviting fire back into ecosystems in the western United States would support the goal of reestablishing resilience at scale. The importance of maintenance work would remain because of the safety, human health, and economic considerations at the wildland–urban interface, in particular. In Hessburg's view, reintroducing resilience at scale would mean reducing the need for aggressive fire suppression because fire would do some of work of ecosystem management, allowing for strategic management and suppression work around human settlements. Fox noted the importance of increasing capacity and resources for nature-based solutions for mitigation that have the co-benefits of clean air and water and protecting species at risk. Given the large U.S. National Wildlife Refuge system and amount of the nation's carbon stocks in Alaska, Fox noted the opportunity to expand capacity to support carbon management. Murulitharan reflected on the establishment of a regional policy framework in Southeast Asia over the last 20 years to address wildfire incidence in the region. This regional, legally binding policy promotes cooperation between countries to fight and mitigate wildfires. While countries have considered or implemented laws on transboundary pollution that seek to prosecute perpetrators responsible for wildfires, Murulitharan argued for investing in a strong scientific framework to support regional cooperation and management.

Closing Thoughts

Climate change is fundamentally changing ecosystems and their fire conditions. The 2023 fire season, still active during the workshop, highlighted the urgency of developing and implementing solutions to address wildland fires. While the focus of the workshop was on improving understanding to ultimately reduce greenhouse gas (GHG) emissions from wildland fires, many workshop participants emphasized the importance of balancing mitigation efforts that are climate effective, socially inclusive, and ecologically appropriate. Key to the consideration of how direct and indirect (i.e., climate change-driven) human influences on GHG emissions from wildland fires may impact efforts to reach net-zero emissions, several speakers highlighted that the reduction of GHG emissions from anthropogenic sources could be critical for slowing the impacts from climate change on fires and ecosystems.

A key theme highlighted throughout the workshop was the importance of not only learning from current and historic practices of Indigenous peoples, but also centering Indigenous voices and leadership in all stages of fire management. Many participants emphasized that engaging local communities who live in and manage these systems and benefit from land use is a critical part of identifying and implementing appropriate intervention strategies. Colonial policies of fire suppression have disrupted historical land use and cultural practices, and today, communities largely do not have the opportunity to reintroduce fire to the landscape and steward their lands. Discussions highlighted opportunities to shift the current regime of fire management to allow fires of choice and reduce GHG emissions, including reintroducing cultural burning and cultural land management practices.

Workshop discussions centered on ecosystems particularly impacted by climate and land use changes where historical fire regimes and the carbon balance have been disrupted. Discussions were organized about three global biomes—temperate, Arctic/boreal, and tropical—with the recognition that ecosystems are heterogeneous and diverse in their ecological conditions, culture, and landscape resilience.

In temperate forest systems, both direct and indirect (i.e., climate change) human-driven changes to the fire regime are threatening carbon stocks, or carbon carrying capacity. Changes in the frequency or intensity of fires have impacted the ability of these systems to regenerate or regrow vegetation that historically served a self-regulating function. In the western U.S. dry, temperate ecosystems in particular, a century of fire suppression has led to dense forests that have likely exceeded their carbon carrying capacity with an elevated risk of high-severity fire. Today's conditions are very different from historic landscapes where there were lower-density forests and carbon stocks that had frequent lower-severity fire. Fuel and landscape management solutions will likely be important going forward, including cultural burning, prescribed burning, active management of wildfire, and mechanical fuel reduction.

Increases in the frequency and severity of wildland fires due to climate change in the Arctic and boreal regions is threatening the large amounts of carbon that have been stored for decades or centuries as permafrost. Burning in boreal systems is dominated by combustion of belowground carbon. As fire severity in these systems increases, it will decrease the net primary productivity of boreal forests, thin the organic layer on the ground, warm soils, and lead to faster decomposition and turnover of biomass—a cycle that will continue and could lead to permanent permafrost (carbon) loss after fire.

Arctic and boreal systems are warming more rapidly than global average temperatures; however, reducing emissions from future fires is challenging because these are very large areas with low population density and limited infrastructure. Within the boreal, there are large regional differences in flammability, carbons stocks (e.g., permafrost, peat), and fire risk. Responding to massive and accelerating changes in the boreal may involve large investments in people and response mechanisms. Global science and policy collaboration and funding may be needed to substantially reduce emissions across the Arctic and boreal region. Solutions in the boreal may involve a mix of approaches, including targeted suppression of early-stage fires in carbon and Yedoma permafrost–rich coniferous habitats, cultural burning, post-fire seeding to accelerate recovery, and in some geographies, stopping the suppression of broadleaf species to reduce fire spread, which may include economic incentives to utilize broadleaf species as part of the bioeconomy. However, there are trade-offs in the reduction of conifer habitats that are important for some endangered species and for trapline users.

In tropical peatland systems, rapid large-scale land use change—from agriculture and land clearing—has drained peatlands making them more vulnerable to fire. Natural climate variability (e.g., El Niño–Southern Oscillation) can amplify fire activity in these regions and increase GHG emissions. In tropical peatland systems where human activities have disrupted the ecosystems, water management to re-wet and maintain higher water tables in peatlands and land management that protects systems that are burning frequently for agriculture could reduce GHG emissions and protect ecosystems. However, fire in many tropical and subtropical ecosystems is essential for food security, so it is important to balance livelihood security with fire prevention. Tropical forests are also at increasing risk due to changes in the three drivers of fire in these systems: fuels, ignitions, and climate. Reducing deforestation could be the most effective strategy to manage wildfire risk.

In Australia, where there is a mix of ecosystem types, low-intensity burning early in the fire season can reduce high-intensity, late-season fires and their emissions in the northern tropical savannas. On the other hand, in temperate forests in Australia, ignition management together with strategic applications of prescribed burning, could reduce emissions from wildfires.

To implement the suite of fire and land management strategies discussed throughout the workshop, several participants highlighted that investments in resources—including for workforce and capacity—would be useful. Speakers noted the difference between cost-effective (i.e., cost per ton of mitigation) and low-cost solutions, and pointed to an increase in resource allocation that may be important to match the scale of mitigation required.

However, the use of carbon markets to finance fire management could disincentivize overall reductions in emissions.

Several speakers identified opportunities to advance observational and modeling tools to improve understanding of current and future fires and their associated GHG emissions. Given the complexity of fuels across all types of landscapes, some participants highlighted a goal to better characterize the total mass, physical structure, and chemical and water dynamics of fuels. Relatedly, better characterization of fire behavior and fuel consumption—including the spatial variation and moisture content—could improve both observation- and model-based GHG emission estimates. From a carbon emissions perspective, there is a gap in accurately characterizing prescribed, small, and cooler (e.g., surface, nighttime, peat) fires, and emissions from duff, peatlands, and permafrost. Continuous, consistent satellite observations also have a role to play, and high-resolution geostationary observations and trace gas retrievals are an opportunity to improve understanding at higher spatial resolutions.

In addition to observations, there are diverse mechanisms that represent fire ignition, spread, suppression, and extinction in models where decisions are made about the process-level detail to represent fire at large scales. Model representations of vegetation–fire feedbacks are particularly important for predicting future fires. For example, in tropical savanna systems, interactions between fire, carbon dioxide, shrubs, and grass may have a positive feedback on emissions. To build confidence in projections of wildland fires and their emissions, participants discussed the importance of evaluating how well models simulate observed spatial patterns, annual cycles, interannual variability, and long-term trends of fire and land cover. Additionally, accurately simulating succession could be an important test of the ability of dynamic vegetation models to capture fire processes.

Another accounting mechanism important for decision making is the reporting of inventories of national GHG emissions by countries as part of international agreements. This reporting by design focuses on direct human (anthropogenic) emissions and removals. Emissions from wildland fires, if they occur on unmanaged lands or on some managed lands, are not accounted for as part of national inventories; on the other hand, prescribed fire emissions are always accounted for. Countries—including the United States, Canada, and Australia—are working to reduce uncertainties of national wildfire emission estimates, particularly by expanding inventories into managed and unmanaged forests. However, some participants highlighted the importance of being realistic about the limitations of the land sector to increase or even maintain current carbon sinks as climate impacts increase.

The enthusiastic participation of the academic, governmental, private, practitioner, and Indigenous communities in the workshop demonstrated the energy around improving understanding and finding solutions to reduce GHG emissions from wildland fires. Participants discussed opportunities to shift the current regime of management toward carbon-focused management that could increase the resilience of ecosystems to store carbon. This moment of increasing vulnerability to wildfires is a chance to turn attention toward the diverse set of available regionally differentiated, ecosystem-appropriate mitigation strategies.

References

Abatzoglou, J. T., and A. P. Williams. 2016. Impact of anthropogenic climate change on wildfire across western US forests. *Proceedings of the National Academy of Sciences of the United States of America* 113(42):11770-11775. https://doi.org/10.1073/pnas.1607171113.

Abatzoglou, J. T., D. S. Battisti, A. P. Williams, W. D. Hansen, B. J. Harvey, and C. A. Kolden. 2021. Projected increases in western US forest fire despite growing fuel constraints. *Communications Earth & Environment* 2(1):227. https://doi.org/10.1038/s43247-021-00299-0.

Alencar, A., P. Moutinho, V. Arruda, and D. Silvério. 2020. *Amazonas em chamas—O fogo e o desmatamento me 2019 e o que vem em 2020*. Nota técnica no. 3. Brasília: Instituto de Pesquisa Ambiental da Amazônia. https://ipam.org.br/bibliotecas/amazoniaem-chamas-3-o-fogo-e-o-desmatamentoem-2019-e-o-que-ve—-em-2020.

Anderson, L. O., C. Burton, J. B. C. dos Reis, A. C. M. Pessôa, P. Bett, N. S. Carvalho, C. H. L. Silva, Jr., K. Williams, G. Selaya, D. Armenteras, B. A. Bilbao, H. A. M. Xaud, R. Rivera-Lombardi, J. Ferreira, L. E. O. C. Aragão, C. D. Jones, and A. J. Wiltshire. 2022. An alert system for seasonal fire probability forecast for South American protected areas. *Climate Resilience and Sustainability* 1(1):e19. https://doi.org/10.1002/cli2.19.

Archibald, S., C. E. R. Lehmann, J. L. Gómez-Dans, and R. A. Bradstock. 2013. Defining pyromes and global syndromes of fire regimes. *Proceedings of the National Academy of Sciences of the United States of America* 110(16):6442-6447. https://doi.org/10.1073/pnas.1211466110.

Balch, J. K., P. M. Brando, D. C. Nepstad, M. T. Coe, D. Silvério, T. J. Massad, E. A. Davidson, P. Lefebvre, C. Oliveira-Santos, W. Rocha, R. T. S. Cury, A. Parsons, and K. S. Carvalho. 2015. The susceptibility of southeastern Amazon forests to fire: Insights from a large-scale burn experiment. *BioScience* 65(9):893-905. https://doi.org/10.1093/biosci/biv106.

Bernier, P. Y., S. Gauthier, P.-O. Jean, F. Manka, Y. Boulanger, A. Beaudoin, and L. Guindon. 2016. Mapping local effects of forest properties on fire risk across Canada. *Forests* 7(8):157. https://doi.org/10.3390/f7080157.

Bilbao, B., A. Leal, C. Méndez, and M. D. Delgado-Cartay. 2009. The role of fire in the vegetation dynamics of upland savannas of the Venezuelan Guayana. In *Tropical Fire Ecology: Climate Change, Land Use, and Ecosystem Dynamics*, M. A. Cochrane, ed. Berlin, Heidelberg: Springer Berlin Heidelberg, pp. 451-480.

Bilbao, B. A., A. V. Leal, and C. L. Méndez. 2010. Indigenous use of fire and forest loss in Canaima National Park, Venezuela. Assessment of and tools for alternative strategies of fire management in Pemón Indigenous lands. *Human Ecology* 38(5):663-673. https://doi.org/10.1007/s10745-010-9344-0.

Bilbao, B., J. Rosales, S. Marin, A. Millan, R. Salazar-Gascon, H. Chani, F. Pérez, A. Leal, C. Méndez, D. Delgado-Cartay, M. Márquez, M. Alvarado, E. Deza, Z. Hasmy, F. Lambos, I. Lanz, R. Machuca, M. Parra, E. P. P. Castellanos, G. P. Nava, F. Reyes, D. Rodriguez, H. Manuel, H. M. Rodriguez Salcedo, B. Sanchez, and E. Zambrano. 2017. Chureta ru to pomupök integration of indigenous and ecological knowledge for the restoration of degraded environments. In *Beyond Restoration Ecology: Social Perspectives in Latin America and the Caribbean,* E. Ceccon and D. Perez, eds. Buenos Aires: Vazquez Mazzini Editores, pp. 331-353.

Bilbao, B. A., A. Millán, H. Vessuri, J. Mistry, R. Salazar-Gascón, and R. Gómez. 2021. To burn or not to burn? The history behind the construction of a new paradigm of fire management in Venezuela through interculturality. *Biodiversidade Brasileira* 11(2):99-127. https://doi.org/10.37002/biodiversidadebrasileira.v11i2.

Bilbao, B. A., A. Millán, M. L. Matany, J. Mistry, R. Gómez-Martínez, R. Rivera-Lombardi, C. Méndez-Vallejo, E. León, J. G. Biskis Ardila, G. Gutiérrez, E. G. León, and B. Ancidey. 2022. An intercultural vision for integrated fire management in Venezuela. *Tropical Forest Issues* 61:39-46. https://doi.org/10.55515/CNUU7417.

Boisramé, G., S. Thompson, B. Collins, and S. Stephens. 2017. Managed wildfire effects on forest resilience and water in the Sierra Nevada. *Ecosystems* 20(4):717-732. https://doi.org/10.1007/s10021-016-0048-1.

Bond, W. J., F. I. Woodward, and G. F. Midgley. 2005. The global distribution of ecosystems in a world without fire. *New Phytologist* 165: 525-538. https://doi.org/10.1111/j.1469-8137.2004.01252.x.

Bowman, D. M. J. S., G. J. Williamson, J. T. Abatzoglou, C. A. Kolden, M. A. Cochrane, and A. M. S. Smith. 2017. Human exposure and sensitivity to globally extreme wildfire events. *Nature Ecology & Evolution* 1(3):58. https://doi.org/10.1038/s41559-016-0058.

Bowman, D. M. J. S., G. J. Williamson, M. Ndalila, S. H. Roxburgh, S. Suitor, and R. J. Keenan. 2023. Wildfire national carbon accounting: How natural and anthropogenic landscape fires emissions are treated in the 2020 Australian government greenhouse gas accounts report to the UNFCCC. *Carbon Balance and Management* 18(1):14. https://doi.org/10.1186/s13021-023-00231-3.

Bradstock, R. A., G. J. Cary, I. Davies, D. B. Lindenmayer, O. F. Price, and R. J. Williams. 2012. Wildfires, fuel treatment and risk mitigation in Australian eucalypt forests: Insights from landscape-scale simulation. *Journal of Environmental Management* 105:66-75. https://doi.org/10.1016/j.jenvman.2012.03.050.

Burton, C., S. Lampe, D. Kelley, W. Thiery, S. Hantson, N. Christidis, L. Gudmundson, M. Forrest, E. Burke, J. Chang, H. Huang, A. Ito, S. Kou-Giesbrecht, G. Lasslop, W. Li, L. Nieradzik, F. Li, Y. Chen, J. Randerson, C. Reyer, and M. Mengel. 2023. Global burned area increasingly explained by climate change. *Research Square,* preprint. https://doi.org/10.21203/rs.3.rs-3168150/v1.

Calkin, D. E., K. Barrett, J. D. Cohen, M. A. Finney, S. J. Pyne, and S. L. Quarles. 2023. Wildland-urban fire disasters aren't actually a wildfire problem. *Proceedings of the National Academy of Sciences of the United States of America* 120(51):e2315797120. https://doi.org/10.1073/pnas.2315797120.

Castanho, A. D., M. T. Coe, P. Brando, M. Macedo, A. Baccini, W. Walker, and E. M. Andrade. 2020. Potential shifts in the aboveground biomass and physiognomy of a seasonally dry tropical forest in a changing climate. *Environmental Research Letters* 15(3):034053. https://doi.org/10.1088/1748-9326/ab7394.

Chapin, F. S., S. R. Carpenter, G. P. Kofinas, C. Folke, N. Abel, W. C. Clark, P. Olsson, D. M. S. Smith, B. Walker, O. R. Young, F. Berkes, R. Biggs, J. M. Grove, R. L. Naylor, E. Pinkerton, W. Steffen, and F. J. Swanson. 2010. Ecosystem stewardship: Sustainability strategies for a rapidly changing planet. *Trends in Ecology & Evolution* 25(4):241-249. https://doi.org/10.1016/j.tree.2009.10.008.

Christianson, A. C., C. R. Sutherland, F. Moola, N. Gonzalez Bautista, D. Young, and H. MacDonald. 2022. Centering Indigenous voices: The role of fire in the boreal forest of North America. *Current Forestry Reports* 8(3):257-276. https://doi.org/10.1007/s40725-022-00168-9.

Collins, B. M., J. D. Miller, A. E. Thode, M. Kelly, J. W. van Wagtendonk, and S. L. Stephens. 2009. Interactions among wildland fires in a long-established Sierra Nevada natural fire area. *Ecosystems* 12:114-128. https://doi.org/10.1007/s10021-008-9211-7.

Crippa, M., D. Guizzardi, M. Muntean, E. Schaaf, E. Solazzo, F. Monforti-Ferrario, J. G. J. Olivier, and E. Vignati. 2020. *Fossil CO_2 and GHG Emissions of All World Countries: 2020 Report*. Luxemberg: Publications Office of the European Union.

Cummins, K., J. Noble, J. M. Varner, K. M. Robertson, J. K. Hiers, H. K. Nowell, and E. Simonson. 2023. The southeastern US prescribed fire permit database: Hot spots and hot moments in prescribed fire across the southeastern USA. *Fire* 6(10):372. https://doi.org/10.3390/fire6100372.

Dargie, G., S. Lewis, I. Lawson, E. T. A. Mitchard, S. E. Page, Y. E. Bocko, and S. A. Ifo. 2017. Age, extent and carbon storage of the central Congo Basin peatland complex. *Nature* 542:86–90. https://doi.org/10.1038/nature21048.

Domke, G. M., B. F. Walters, C. L. Giebink, E. J. Greenfield, J. E. Smith, M. C. Nichols, J. A. Knott, S. M. Ogle, J. W. Coulston, and J. Steller. 2023. *Greenhouse Gas Emissions and Removals from Forest Land, Woodlands, Urban Trees, and Harvested Wood Products in the United States, 1990-2021*. U.S. Department of Agriculture, Forest Service. http://dx.doi.org/10.2737/WO-RB-101.

Edwards, A., R. Archer, P. De Bruyn, J. Evans, B. Lewis, T. Vigilante, S. Whyte, and J., Russell-Smith. 2021. Transforming fire management in northern Australia through successful implementation of savanna burning emissions reductions projects. *Journal of Environmental Management* 290:112568. https://doi.org/10.1016/j.jenvman.2021.112568.

Environment and Climate Change Canada. 2023. *National Inventory Report 1990–2021: Greenhouse Gas Sources and Sinks in Canada*. https://unfccc.int/documents/627833.

EPA (U.S. Environmental Protection Agency). 2023. *Inventory of U.S. Greenhouse Gas Emissions and Sinks: 1990-2021*. EPA 430-R-23-002. Washington, DC: U.S. Environmental Protection Agency. https://www.epa.gov/ghgemissions/inventory-us-greenhouse-gas-emissions-and-sinks-1990-2021.

Erni, S., D. Arseneault, and M.-A. Parisien. 2018. Stand age influence on potential wildfire ignition and spread in the boreal forest of northeastern Canada. *Ecosystems* 21(7):1471-1486. https://doi.org/10.1007/s10021-018-0235-3.

Forkel, M., N. Andela, S. P. Harrison, G. Lasslop, M. van Marle, E. Chuvieco, W. Dorigo, M. Forrest, S. Hantson, A. Heil, F. Li, J. Melton, S. Sitch, C. Yue, and A. Arneth. 2019. Emergent relationships with respect to burned area in global satellite observations and fire-enabled vegetation models. *Biogeosciences* 16(1):57-76. https://doi.org/10.5194/bg-16-57-2019.

French, N. H., and A. T. Hudak. 2023. Biomass burning fuel consumption and emissions for air quality. In *Landscape Fire, Smoke, and Health: Linking Biomass Burning Emissions to Human Well-Being*, T. V. Loboda, N. H. French, and R. C. Puett, eds. Hoboken, NJ: Wiley.

Gale, M. G. and G. J. Cary. 2021. Stand boundary effects on obligate seeding Eucalyptus delegatensis regeneration and fuel dynamics following high and low severity fire: Implications for species resilience to recurrent fire. *Austral Ecology* 46(5):802-817. https://doi.org/10.1111/aec.13024.

Genet, H., Y. He, Z. Lyu, A. D. McGuire, Q. Zhuang, J. Clein, D. D'Amore, A. Bennett, A. Breen, F. Biles, E. S. Euskirchen, K. Johnson, T. Kurkowski, S. Schroder, N. Pastick, T. S. Rupp, B. Wylie, Y. Zhang, X. Zhou, and Z. Zhu. 2018. The role of driving factors in historical and projected carbon dynamics of upland ecosystems in Alaska. *Ecological Applications* 28(1):5-27. https://doi.org/https://doi.org/10.1002/eap.1641.

Gibson, C. M., L. E. Chasmer, D. K. Thompson, W. L. Quinton, M. D. Flannigan, and D. Olefeldt. 2018. Wildfire as a major driver of recent permafrost thaw in boreal peatlands. *Nature Communications* 9(1):3041. https://doi.org/10.1038/s41467-018-05457-1.

Goodwin, M. J., H. S. Zald, M. P. North, and M. D. Hurteau. 2021. Climate-driven tree mortality and fuel aridity increase wildfire's potential heat flux. *Geophysical Research Letters* 48(24):e2021GL094954. https://doi.org/10.1029/2021GL094954.

Hagmann, R. K., P. F. Hessburg, S. J. Prichard, N. A. Povak, P. M. Brown, P. Z. Fulé, R. E. Keane, E. E. Knapp, J. M. Lydersen, K. L. Metlen, M. J. Reilly, A. J. Sánchez Meador, S. L. Stephens, J. T. Stevens, A. H. Taylor, L. L. Yocom, M. A. Battaglia, D. J. Churchill, L. D. Daniels, D. A. Falk, P. Henson, J. D. Johnston, M. A. Krawchuk, C. R. Levine, G. W. Meigs, A. G. Merschel, M. P. North, H. D. Safford, T. W. Swetnam, and A. E. M. Waltz. 2021. Evidence for widespread changes in the structure, composition, and fire regimes of western North American forests. *Ecological Applications* 31(8):e02431. https://doi.org/10.1002/eap.2431.

Hansen, W. D., M. A. Krawchuk, A. T. Trugman, and A. P. Williams. 2022. The Dynamic Temperate and Boreal Fire and Forest-Ecosystem Simulator (DYNAFFOREST): Development and evaluation. *Environmental Modelling & Software* 156:105473. https://doi.org/10.1016/j.envsoft.2022.105473.

Hantson, S., A. Arneth, S. P. Harrison, D. I. Kelley, I. C. Prentice, S. S. Rabin, S. Archibald, F. Mouillot, S. R. Arnold, P. Artaxo, D. Bachelet, P. Ciais, M. Forrest, P. Friedlingstein, T. Hickler, J. O. Kaplan, S. Kloster, W. Knorr, G. Lasslop, F. Li, S. Mangeon, J. R. Melton, A. Meyn, S. Sitch, A. Spessa, G. R. van der Werf, A. Voulgarakis, and C. Yue. 2016. The status and challenge of global fire modelling. *Biogeosciences* 13(11):3359-3375. https://doi.org/10.5194/bg-13-3359-2016.

Hantson, S., D. I. Kelley, A. Arneth, S. P. Harrison, S. Archibald, D. Bachelet, M. Forrest, T. Hickler, G. Lasslop, F. Li, S. Mangeon, J. R. Melton, L. Nieradzik, S. S. Rabin, I. C. Prentice, T. Sheehan, S. Sitch, L. Teckentrup, A. Voulgarakis, and C. Yue. 2020. Quantitative assessment of fire and vegetation properties in simulations with fire-enabled vegetation models from the Fire Model Intercomparison Project. *Geoscientific Model Development* 13(7):3299-3318. https://doi.org/10.5194/gmd-13-3299-2020.

Harris, R. M. B., T. A. Remenyi, G. J. Williamson, N. L. Bindoff, and D. M. J. S. Bowman. 2016. Climate–vegetation–fire interactions and feedbacks: Trivial detail or major barrier to projecting the future of the Earth system? *WIREs Climate Change* 7(6):910-931. https://doi.org/10.1002/wcc.428.

Hessburg, P. F., T. A. Spies, D. A. Perry, C. N. Skinner, A. H. Taylor, P. M. Brown, S. L. Stephens, A. J. Larson, D. J. Churchill, N. A. Povak, P. H. Singleton, B. McComb, W. J. Zielinski, B. M. Collins, R. B. Salter, J. J. Keane, J. F. Franklin, and G. Riegel. 2016.

Tamm Review: Management of mixed-severity fire regime forests in Oregon, Washington, and Northern California. *Forest Ecology and Management* 366:221-250. https://doi.org/10.1016/j.foreco.2016.01.034.

Hessburg, P. F., C. L. Miller, S. A. Parks, N. A. Povak, A. H. Taylor, P. E. Higuera, S. J. Prichard, M. P. North, B. M. Collins, M. D. Hurteau, A. J. Larson, C. D. Allen, S. L. Stephens, H. Rivera-Huerta, C. S. Stevens-Rumann, L. D. Daniels, Z. Gedalof, R. W. Gray, V. R. Kane, D. J. Churchill, R. K. Hagmann, T. A. Spies, C. A. Cansler, R. T. Belote, T. T. Veblen, M. A. Battaglia, C. Hoffman, C. N. Skinner, H. D. Safford, and R. B. Salter. 2019. Climate, environment, and disturbance history govern resilience of western North American forests. *Frontiers in Ecology and Evolution* 7. https://doi.org/10.3389/fevo.2019.00239.

Hu, Y., N. Fernandez-Anez, T. E. L. Smith, and G. Rein. 2018. Review of emissions from smouldering peat fires and their contribution to regional haze episodes. *International Journal of Wildland Fire* 27(5):293-312. https://doi.org/10.1071/WF17084.

Hurteau, M. D. 2013. Effects of wildland fire management on forest carbon stores. In *Land use and the carbon cycle: Advances in integrated science, management, and policy*. D.G. Brown, D.T. Robinson, N.H.F. French, and B.C. Reed, eds. Cambridge, UK: Cambridge University Press. https://doi.org/10.1017/CBO9780511894824.

IPCC (Intergovernmental Panel on Climate Change). 2006. *2006 IPCC Guidelines for National Greenhouse Gas Inventories*. S. Eggelston, L. Buendia, K. Miwa, T. Ngara, and K. Tanabe, eds. Hayama, Kanagawa, Japan: Institute for Global Environmental Strategies. https://www.ipcc-nggip.iges.or.jp/public/2006gl/index.html.

IPCC. 2021. *Climate Change 2021: The Physical Science Basis. Contribution of Working Group I to the Sixth Assessment Report of the Intergovernmental Panel on Climate Change*, V. Masson-Delmotte, P. Zhai, A. Pirani, S. L. Connors, C. Péan, S. Berger, N. Caud, Y. Chen, L. Goldfarb, M. I. Gomis, M. Huang, K. Leitzell, E. Lonnoy, J. B. R. Matthews, T. K. Maycock, T. Waterfield, O. Yelekçi, R. Yu, and B. Zhou, eds. Cambridge, UK, and New York, NY: Cambridge University Press. https://doi.org/10.1017/9781009157896.

Jafarov, E. E., V. E. Romanovsky, H. Genet, A. D. McGuire, and S. S. Marchenko. 2013. The effects of fire on the thermal stability of permafrost in lowland and upland black spruce forests of interior Alaska in a changing climate. *Environmental Research Letters* 8(3):035030. https://doi.org/10.1088/1748-9326/8/3/035030.

Jia, G., E. Shevliakova, P. Artaxo, N. D. Noblet-Ducoudré, R. Houghton, J. House, K. Kitajima, C. Lennard, A. Popp, A. Sirin, R. Sukumar, and L. Verchot. 2019. Land–climate interactions. In *Climate Change and Land: IPCC Special Report on Climate Change, Desertification, Land Degradation, Sustainable Land Management, Food Security, and Greenhouse Gas Fluxes in Terrestrial Ecosystems*, P. R. Shukla, J. Skea, E. C. Buendia, V. Masson-Delmotte, H.-O. Pörtner, D. C. Roberts, P. Zhai, R. Slade, S. Connors, R. van Diemen, M. Ferrat, E. Haughey, S. Luz, S. Neogi, M. Pathak, J. Petzold, J. P. Pereira, P. Vyas, E. Huntley, K. Kissick, M. Belkacemi, and J. Malley, eds. Cambridg, UK: Cambridge University Press.

Johnstone, J. F., T. S. Rupp, M. Olson, and D. Verbyla. 2011. Modeling impacts of fire severity on successional trajectories and future fire behavior in Alaskan boreal forests. *Landscape Ecology* 26(4):487-500. https://doi.org/10.1007/s10980-011-9574-6.

Juang, C. S., A. P. Williams, J. T. Abatzoglou, J. K. Balch, M. D. Hurteau, and M. A. Moritz. 2022. Rapid growth of large forest fires drives the exponential response of annual forest-fire area to aridity in the western United States. *Geophysical Research Letters* 49(5):e2021 GL097131. https://doi.org/10.1029/2021GL097131.

Kaiser, J. W., A. Heil, M. O. Andreae, A. Benedetti, N. Chubarova, L. Jones, J. J. Morcrette, M. Razinger, M. G. Schultz, M. Suttie, and G. R. van der Werf. 2012. Biomass burning emissions estimated with a global fire assimilation system based on observed fire radiative power. *Biogeosciences* 9(1):527-554. https://doi.org/10.5194/bg-9-527-2012.

Kelly, R., M. L. Chipman, P. E. Higuera, I. Stefanova, L. B. Brubaker, and F. S. Hu. 2013. Recent burning of boreal forests exceeds fire regime limits of the past 10,000 years. *Proceedings of the National Academy of Sciences of the United States of America* 110(32):13055-13060. https://doi.org/10.1073/pnas.1305069110.

Kharuk, V. I., M. L. Dvinskaya, S. T. Im, A. S. Golyukov, and K. T. Smith. 2022. Wildfires in the Siberian Arctic. *Fire* 5(4):106. https://doi.org/10.3390/fire5040106.

Kloster, S., and G. Lasslop. 2017. Historical and future fire occurrence (1850 to 2100) simulated in CMIP5 Earth System Models. *Global and Planetary Change* 150:58-69. https://doi.org/10.1016/j.gloplacha.2016.12.017.

Koch, A., and J. O. Kaplan. 2022. Tropical forest restoration under future climate change. *Nature Climate Change* 12(3):279-283. https://doi.org/10.1038/s41558-022-01289-6.

Koplitz, S. N., L. J. Mickley, M. E. Marlier, J. J. Buonocore, P. S. Kim, T. Liu, M. P. Sulprizio, R. S. DeFries, D. J. Jacob, J. Schwartz, M. Pongsiri, and S. S. Myers. 2016. Public health impacts of the severe haze in Equatorial Asia in September–October 2015: Demonstration of a new framework for informing fire management strategies to reduce downwind smoke exposure. *Environmental Research Letters* 11(9):094023. https://doi.org/10.1088/1748-9326/11/9/094023.

Kreye, J. K., J. M. Varner, and L. N. Kobziar. 2020. Long-duration soil heating resulting from forest floor duff smoldering in longleaf pine ecosystems. *Forest Science* 66(3):291-303. https://doi.org/10.1093/forsci/fxz089.

Krofcheck, D. J., C. C. Remy, A. R. Keyser, and M. D. Hurteau. 2019. Optimizing forest management stabilizes carbon under projected climate and wildfires. *Journal of Geophysical Research: Biogeosciences* 124(10):3075-3087. https://doi.org/10.1029/2019JG005206.

Kruid, S., M. N. Macedo, S. R. Gorelik, W. Walker, P. Moutinho, P. M. Brando, A. Castanho, A. Alencar, A. Baccini, and M. T. Coe. 2021. Beyond deforestation: Carbon emissions from land grabbing and forest degradation in the Brazilian Amazon. *Frontiers in Forests and Global Change* 4. https://doi.org/10.3389/ffgc.2021.645282.

Lake, F. K., and A. C. Christianson. 2019. Indigenous fire stewardship. In *Encyclopedia of Wildfires and Wildland-Urban Interface (WUI) Fires*, S. L. Manzello, ed. Cham: Springer International.

Landry, J. S., and H. D. Matthews. 2016. Non-deforestation fire vs. fossil fuel combustion: The source of CO_2 emissions affects the global carbon cycle and climate responses. *Biogeosciences* 13(7):2137-2149. https://doi.org/10.5194/bg-13-2137-2016.

Larkin, N. K., T. M. Strand, S. A. Drury, S. M. Raffuse, R. C. Solomon, S. M. O'Neill, N. Wheeler, S. Huang, M. Roring, and H. R. Hafner. 2012. Phase 1 of the Smoke and Emissions Model Intercomparison Project (SEMIP): Creation of SEMIP and evaluation of current models. Final Report to the Joint Fire Science Program Project #08-1-6-10.

Leenhouts, B. 1998. Assessment of biomass burning in the conterminous United States. *Conservation Ecology* 2(1).

Lewis, H. T., and T. A. Ferguson. 1988. Yards, corridors, and mosaics: How to burn a boreal forest. *Human Ecology* 16(1):57-77. http://www.jstor.org/stable/4602869.

Maclean, K., D. L. Hankins, A. C. Christianson, I. Oliveras, B. A. Bilbao, O. Costello, E. R. Langer, and C. J. Robinson. 2023. Revitalising Indigenous cultural fire practice: Benefits and partnerships. *Trends in Ecology & Evolution*. https://doi.org/10.1016/j.tree.2023.07.001.

Melvin, M. A. 2022. *2021 National Prescribed Fire Use Survey Report*. Technical Report 01-22. Washington, DC: National Association of State Foresters.

Melvin, A. M., M. C. Mack, J. F. Johnstone, A. David McGuire, H. Genet, and E. A. G. Schuur. 2015. Differences in ecosystem carbon distribution and nutrient cycling linked to forest tree species composition in a mid-successional boreal forest. *Ecosystems* 18(8):1472-1488. https://doi.org/10.1007/s10021-015-9912-7.

Miettinen, J., C. Shi, and S. C. Liew. 2016. Land cover distribution in the peatlands of Peninsular Malaysia, Sumatra and Borneo in 2015 with changes since 1990. *Global Ecology and Conservation* 6:67-78. https://doi.org/10.1016/j.gecco.2016.02.004.

Miettinen, J., A. Hooijer, R. Vernimmen, S. C. Liew, and S. E. Page. 2017. From carbon sink to carbon source: Extensive peat oxidation in insular Southeast Asia since 1990. *Environmental Research Letters* 12(2):024014. https://doi.org/10.1088/1748-9326/aa5b6f.

Nguyen, H. M., J. He, and M. J. Wooster. 2023. Biomass burning CO, PM and fuel consumption per unit burned area estimates derived across Africa using geostationary SEVIRI fire radiative power and Sentinel-5P CO data. *Atmospheric Chemistry and Physics* 23(3):2089-2118. https://doi.org/10.5194/acp-23-2089-2023.

Ogle, S. M., G. Domke, W. A. Kurz, M. T. Rocha, T. Huffman, A. Swan, J. E. Smith, C. Woodall, and T. Krug. 2018. Delineating managed land for reporting national greenhouse gas emissions and removals to the United Nations Framework Convention on Climate Change. *Carbon Balance and Management* 13(1):9. https://doi.org/10.1186/s13021-018-0095-3.

Page, S. E., J. O. Rieley, and C. J. Banks. 2011. Global and regional importance of the tropical peatland carbon pool. *Global Change Biology* 17(2):798-818. https://doi.org/10.1111/j.1365-2486.2010.02279.x.

Parks, S. A., C. Miller, M.-A. Parisien, L. M. Holsinger, S. Z. Dobrowski, and J. Abatzoglou. 2015. Wildland fire deficit and surplus in the western United States, 1984–2012. *Ecosphere* 6(12):275. http://dx.doi.org/10.1890/ES15-00294.1.

Phillips, C. A., B. M. Rogers, M. Elder, S. Cooperdock, M. Moubarak, J. T. Randerson, and P. C. Frumhoff. 2022. Escalating carbon emissions from North American boreal forest wildfires and the climate mitigation potential of fire management. *Science Advances* 8(17):e abl7161. https://doi.org/10.1126/sciadv.abl7161.

Povak, N. A., P. F. Hessburg, R. B. Salter, R. W. Gray, and S. J. Prichard. 2023. System-level feedbacks of active fire regimes in large landscapes. *Fire Ecology* 19:45. https://doi.org/10.1186/s42408-023-00197-0.

Prentice, I. C., A. Bondeau, W. Cramer, S. P. Harrison, T. Hickler, W. Lucht, S. Sitch, B. Smith, and M. T. Sykes. 2007. Dynamic global vegetation modeling: Quantifying terrestrial ecosystem responses to large-scale environmental change. In *Terrestrial Ecosystems in a*

Changing World, J. G. Canadell, D. E. Pataki, and L. F. Pitelka, eds. Berlin, Heidelberg: Springer Berlin Heidelberg, pp. 175-192.
Prichard, S. J., C. S. Stevens-Rumann, and P. F. Hessburg. 2017. Tamm review: Shifting global fire regimes: Lessons from reburns and research needs. *Forest Ecology and Management* 396:217-233. https://doi.org/10.1016/j.foreco.2017.03.035.
Prichard, S. J., E. M. Rowell, A. T. Hudak, R. E. Keane, E. L. Loudermilk, D. C. Lutes, R. D. Ottmar, L. M. Chappell, J. A. Hall, and B. S. Hornsby. 2022. Fuels and consumption. In *Wildland Fire Smoke in the United States: A Scientific Assessment*, D. L. Peterson, S. M. McCaffrey, and T. Patel-Weynand, eds. Cham: Springer International, pp. 11-49.
Pyne, S. J. 2007. *Awful Splendour: A Fire History of Canada*. Vancouver, British Columbia, Canada: UBC Press.
Rabin, S. S., J. R. Melton, G. Lasslop, D. Bachelet, M. Forrest, S. Hantson, J. O. Kaplan, F. Li, S. Mangeon, D. S. Ward, C. Yue, V. K. Arora, T. Hickler, S. Kloster, W. Knorr, L. Nieradzik, A. Spessa, G. A. Folberth, T. Sheehan, A. Voulgarakis, D. I. Kelley, I. C. Prentice, S. Sitch, S. Harrison, and A. Arneth. 2017. The Fire Modeling Intercomparison Project (FireMIP), phase 1: Experimental and analytical protocols with detailed model descriptions. *Geoscientific Model Development* 10(3):1175-1197. https://doi.org/10.5194/gmd-10-1175-2017.
Roos, C. I., C. H. Guiterman, E. Q. Margolis, T. W. Swetnam, N. C. Laluk, K. F. Thompson, C. Toya, C. A. Farris, P. Z. Fulé, J. M. Iniguez, J. M. Kaib, C. D. O'Connor, and L. Whitehair. 2022. Indigenous fire management and cross-scale fire-climate relationships in the Southwest United States from 1500 to 1900 CE. *Science Advances* 8(49):eabq3221. https://doi.org/10.1126/sciadv.abq3221.
Russell-Smith, J., G. D. Cook, P. M. Cooke, A. C. Edwards, M. Lendrum, C. Meyer, and P. J. Whitehead. 2013. Managing fire regimes in north Australian savannas: Applying Aboriginal approaches to contemporary global problems. *Frontiers in Ecology and the Environment* 11:e55-e63. https://doi.org/10.1890/120251.
Safford, H. D., A. K. Paulson, Z. L. Steel, D. J. N. Young, and R. B. Wayman. 2022. The 2020 California fire season: A year like no other, a return to the past or a harbinger of the future? *Global Ecology and Biogeography* 31(10):2005-2025. https://doi.org/10.1111/geb.13498.
Sánchez-López, N., A. T. Hudak, L. Boschetti, C. A. Silva, K. Robertson, E. L. Loudermilk, B. C. Bright, M. A. Callaham, and M. K. Taylor. 2023. A spatially explicit model of tree leaf litter accumulation in fire maintained longleaf pine forests of the southeastern US. *Ecological Modelling* 481:110369. https://doi.org/10.1016/j.ecolmodel.2023.110369.
Sayre, R., D. Karagulle, C. Frye, T. Boucher, N. H. Wolff, S. Breyer, D. Wright, M. Martin, K. Butler, K. Van Graafeiland, and J. Touval. 2020. An assessment of the representation of ecosystems in global protected areas using new maps of world climate regions and world ecosystems. *Global Ecology and Conservation* 21:e00860. https://doi.org/10.1016/j.gecco.2019.e00860.
Silvério, D. V., R. S. Oliveira, B. M. Flores, P. M. Brando, H. K. Almada, M. T. Furtado, F. G. Moreira, M. Heckenberger, K. Y. Ono, and M. N. Macedo. 2022. Intensification of fire regimes and forest loss in the Território Indígena do Xingu. *Environmental Research Letters* 17(4):045012. https://doi.org/10.1088/1748-9326/ac5713.
Sperling, S., M. J. Wooster, and B. D. Malamud. 2020. Influence of satellite sensor pixel size and overpass time on undercounting of Cerrado/Savannah landscape-scale fire radiative

power (FRP): An assessment using the MODIS airborne simulator. *Fire* 3(2):11. https://doi.org/10.3390/fire3020011.

Stephens, S. L., R. E. Martin, and N. E. Clinton. 2007. Prehistoric fire area and emissions from California's forests, woodlands, shrublands, and grasslands. *Forest Ecology and Management* 251(3):205-216. https://doi.org/10.1016/j.foreco.2007.06.005.

Stephens, S. L., S. Thompson, G. Boisramé, B. M. Collins, L.C. Ponisio, E. Rakhmatulina, Z. L. Steel, J. T. Stevens, J. W. van Wagtendonk, and K. Wilkin, 2021. Fire, water, and biodiversity in the Sierra Nevada: A possible triple win. *Environmental Research Communications* 3(8):081004. https://doi.org/10.1088/2515-7620/ac17e2.

Swetnam, T. W., J. Farella, C. I. Roos, M. J. Liebmann, D. A. Falk, and C. D. Allen. 2016. Multiscale perspectives of fire, climate and humans in western North America and the Jemez Mountains, USA. *Philosophical Transactions of the Royal Society B: Biological Sciences* 371(1696):20150168. https://doi.org/10.1098/rstb.2015.0168.

Trumbore, S., P. Brando, and H. Hartmann. 2015. Forest health and global change. *Science* 349(6250):814-818. https://doi.org/10.1126/science.aac6759.

Turco, M., J. T. Abatzoglou, S. Herrera, Y. Zhuang, S. Jerez, D. D. Lucas, A. AghaKouchak, and I. Cvijanovic. 2023. Anthropogenic climate change impacts exacerbate summer forest fires in California. *Proceedings of the National Academy of Sciences of the United States* 120(25):e2213815120. https://doi.org/10.1073/pnas.2213815120.

UNEP (United Nations Environment Programme). 2022. *Spreading like Wildfire: The Rising Threat of Extraordinary Landscape Fires*. Nairobi. https://www.unep.org/resources/report/spreading-wildfire-rising-threat-extraordinary-landscape-fires.

van der Werf, G. R., J. T. Randerson, L. Giglio, G. J. Collatz, M. Mu, P. S. Kasibhatla, D. C. Morton, R. S. DeFries, Y. Jin, and T. T. van Leeuwen. 2010. Global fire emissions and the contribution of deforestation, savanna, forest, agricultural, and peat fires (1997–2009). *Atmospheric Chemistry and Physics* 10(23):11707-11735. https://doi.org/10.5194/acp-10-11707-2010.

Vanderhoof, M. K., T. J. Hawbaker, C. Teske, A. Ku, J. Noble, and J. Picotte. 2021. Mapping wetland burned area from Sentinel-2 across the southeastern United States and its contributions relative to Landsat-8 (2016–2019). *Fire* 4(3):52. https://doi.org/10.3390/fire4030052.

Veraverbeke, S., C. J. F. Delcourt, E. Kukavskaya, M. Mack, X. Walker, T. Hessilt, B. Rogers, and R. C. Scholten. 2021. Direct and longer-term carbon emissions from Arctic-boreal fires: A short review of recent advances. *Current Opinion in Environmental Science & Health* 23:100277. https://doi.org/10.1016/j.coesh.2021.100277.

Walker, X. J., J. L. Baltzer, S. G. Cumming, N. J. Day, C. Ebert, S. Goetz, J. F. Johnstone, S. Potter, B. M. Rogers, E. A. G. Schuur, M. R. Turetsky, and M. C. Mack. 2019. Increasing wildfires threaten historic carbon sink of boreal forest soils. *Nature* 572(7770):520-523. https://doi.org/10.1038/s41586-019-1474-y.

Wang, J. A., J. T. Randerson, M. L. Goulden, C. A. Knight, and J. J. Battles. 2022. Losses of tree cover in California driven by increasing fire disturbance and climate stress. *AGU Advances* 3(4):e2021AV000654. https://doi.org/10.1029/2021AV000654.

Westerling, A. L., Hidalgo, G. H., Cayan, R. D., and T. W. Swetnam. 2006. Warming and earlier spring increase western US forest wildfire activity. *Science* 313(5789):940-943. https://doi.org/10.1126/science.1128834.

Wildland Fire Mitigation and Management Commission. 2023. *On Fire: The Report of the Wildland Fire Mitigation and Management Commission*. Washington, DC: US Department of Agriculture. https://www.usda.gov/sites/default/files/documents/wfmmc-final-report-09-2023.pdf.

Wooster, M. J., G. Roberts, G. L. W. Perry, and Y. J. Kaufman. 2005. Retrieval of biomass combustion rates and totals from fire radiative power observations: FRP derivation and calibration relationships between biomass consumption and fire radiative energy release. *Journal of Geophysical Research: Atmospheres* 110(D24). https://doi.org/10.1029/2005JD006318.

Zheng, B., F. Chevallier, P. Ciais, Y. Yin, M. N. Deeter, H. M. Worden, Y. Wang, Q. Zhang, and K. He. 2018a. Rapid decline in carbon monoxide emissions and export from East Asia between years 2005 and 2016. *Environmental Research Letters* 13(4):044007. https://doi.org/10.1088/1748-9326/aab2b3.

Zheng, B., F. Chevallier, P. Ciais, Y. Yin, and Y. Wang. 2018b. On the role of the flaming to smoldering transition in the seasonal cycle of African fire emissions. *Geophysical Research Letters* 45(21):11,998-12,007. https://doi.org/10.1029/2018GL079092.

Zheng, B., P. Ciais, F. Chevallier, E. Chuvieco, Y. Chen, and H. Yang. 2021. Increasing forest fire emissions despite the decline in global burned area. *Science Advances* 7(39):eabh2646. https://doi.org/10.1126/sciadv.abh2646.

Zheng, B., P. Ciais, F. Chevallier, H. Yang, J. G. Canadell, Y. Chen, I. R. van der Velde, I. Aben, E. Chuvieco, S. J. Davis, M. Deeter, C. Hong, Y. Kong, H. Li, H. Li, X. Lin, K. He, and Q. Zhang. 2023. Record-high CO_2 emissions from boreal fires in 2021. *Science* 379(6635):912-917. https://doi.org/10.1126/science.ade0805.

Zubkova, M., M. L. Humber, and L. Giglio. 2023. Is global burned area declining due to cropland expansion? How much do we know based on remotely sensed data? *International Journal of Remote Sensing* 44(4):1132-1150. https://doi.org/10.1080/01431161.2023.2174389.

Appendix A
Statement of Task

The National Academies will plan a workshop addressing greenhouse gas (GHG) emissions from wildland fires, with goals of improving measurements and model projections of emissions, informing management practices that could limit emissions, and considering how changes in these emissions could affect the ability to achieve net-zero GHG emissions targets. In particular, the workshop will consider the likelihood of increasing frequency of megafires in remote regions, where management actions are typically limited, and the potential for these regions to be large GHG emission sources. Workshop discussions will consider the following topics:

- What is current understanding of how changes in wildland fire GHG emissions could affect the ability to achieve net-zero greenhouse gas emissions targets?
- How are GHG emissions from wildland fires measured and estimated? How could these measurements and estimates be improved?
- How are wildland fire GHG emissions projected to change over decade to century timescales? How could these projections be improved?
- How do global climate models incorporate GHG emissions from wildland fires in projections of decadal to centennial climate change?
- What are possible options for wildland fire prevention and forest management practices that could help limit potential GHG emissions? How can Indigenous knowledge on forest and emission management be incorporated in current and future action plans? How could these practices be designed to also address impacts of wildland fires on human health, safety, and ecosystems?

Appendix B
Biographical Sketches of Committee Members

Loretta J. Mickley (*Chair*) is a senior research fellow at Harvard University in the School of Engineering and Applied Sciences. Her research focuses on interactions between climate and atmospheric chemistry. She uses observations and models to investigate the response of air quality to changing climate in the present day and future. She also examines the two-way interactions between atmospheric composition and climate on a wide range of timescales, including the last glacial period, the preindustrial era, and future. Her recent research has focused on the effects of climate change and human activity on fire and smoke exposure in North and South America, Asia, and Australia. Mickley received an M.S. in chemistry and Ph.D. in geophysical sciences from the University of Chicago. She previously served on the National Academies of Sciences, Engineering, and Medicine's Committee to Review the Climate Science Special Report.

Sally Archibald is a professor in the School of Animal Plant and Environmental Sciences at the University of the Witwatersrand, South Africa. She works on understanding the dynamics and biogeography of savanna ecosystems and is the co-principal investigator of the "Future Ecosystems for Africa" project. Archibald's research on global fire regimes has provided new tools for managing fire in conservation areas to promote biodiversity, and her work on savanna ecosystem functioning is contributing toward better definitions of degradation in tropical ecosystems. Archibald is an associate editor for *Ecology Letters* and *Trends in Ecology and Evolution* and serves on the advisory board of the Leverhulme Centre for Wildfires Environment and Society and the Socio-Ecological Observatory for Studying African Woodlands steering committee. Archibald received a Ph.D. in ecology from the University of the Witwatersrand.

Chris (Fern) Ferner is a lecturer at Johns Hopkins University and spent the past decade as the wildland fire solutions specialist and disaster response program information manager at Esri. While at Esri, Ferner supported the use of geographic information systems (GIS) during all aspects of wildland fire including response, recovery, planning, and mitigation. Previously, she served as Esri's wildland fire and public safety technology specialist. Ferner also directly supported agencies and governments around the globe during active incident responses. She has worked professionally in forestry and wildland fire GIS for 20 years. Ferner received a B.S. in biology from St. Andrews Presbyterian College and an M.S. in forestry with an emphasis in GIS and remote sensing from Colorado State University.

Nancy French is a senior scientist with the Michigan Tech Research Institute and adjunct professor of Forest Resources and Environmental Sciences at Michigan Technological

University. She previously worked as a research scientist at the Environmental Research Institute of Michigan. French has expertise in landscape ecology and in using remote sensing and geospatial approaches for the study of wildland fire and fire effects. In her years of research, she has developed approaches to use satellite data to monitor the spatial and temporal patterns of fire, fuels, and smoke. She has experience in working with satellite images from Landsat, MODIS, and SAR imaging systems. Her work includes integrating remote sensing and geospatial products into decision products, including mapping landscape fuels and quantifying wildland fire emissions of carbon and air pollutants. French received a NASA Earth Science Fellowship and has previously served on the North American Carbon Program Science Steering Group, the Commission for Environmental Cooperation Expert Panel on Black Carbon Emissions Estimation Guidelines, and as a contributing author to the Second State of the Carbon Cycle Report. French received a B.S. from Bates College and an M.S. and Ph.D. in natural resources from the University of Michigan, Ann Arbor.

Don Hankins is a professor of geography and planning at California State University, Chico, and field director for the Big Chico Creek Ecological Reserve. Hankins has been involved in various aspects of land stewardship and conservation for a variety of organizations and agencies including federal and Indigenous entities in North America and Australia. His areas of expertise are pyrogeography, water resources, and conservation. Combining his academic and cultural knowledge as a traditional cultural practitioner, he engages in applied research and projects utilizing Indigenous stewardship practices to aid in conservation and resilience. He is engaged in wildland fire research with an emphasis on landscape-scale prescribed and cultural burns, ecocultural restoration, and environmental policy and has published on these topics. Hankins is an advisor to the Indigenous Peoples Burning Network; founder, co-lead, and secretary of the Indigenous Stewardship Network; and an appointed member of the California Wildfire and Forest Resilience Task Force executive committee. Among recent honors, Hankins received recognition with a professional achievement honor as an outstanding teacher-scholar, and as a Grist 50 Fixer. Hankins received a B.S. in wildlife, fish, and conservation biology and a Ph.D. in geography from the University of California, Davis.

He has served in an advisory role in research and policy regarding wildfire emissions and smoke and has made professional presentations and published articles for public awareness on these topics.

Werner Kurz was most recently a senior research scientist at the Canadian Forest Service of Natural Resources Canada in Victoria, BC. He serves as adjunct professor at the University of British Columbia and at Simon Fraser University. He led the development of Canada's National Forest Carbon Monitoring, Accounting and Reporting System and the Wildfire and Carbon Project of the Pacific Institute for Climate Solutions. Kurz's research focuses on carbon dynamics in forests and harvested wood products and the opportunities of the forest sector to contribute to climate change mitigation. He has co-authored eight reports of the Intergovernmental Panel on Climate Change. He is an International Fellow of the Royal Swedish Academy of Agriculture and Forestry. Kurz received a B.Sc. in wood science and technology from the University of Hamburg, Germany, a Ph.D. in forest ecology from the University of British Columbia, and an honorary doctorate from the Swedish Land University.

James Randerson (NAS) is the Ralph J. and Carol M. Cicerone Professor of Earth System Science at the University of California, Irvine (UCI). Prior to joining the faculty at UCI, he was an assistant professor at Caltech from 2000 to 2003. Randerson studies the terrestrial biosphere and the role of fire in the Earth system using high-resolution satellite imagery to identify how fires are changing in response to climate warming and land use intensification. He has conducted field measurements in boreal forests of Siberia and Alaska and temperate forests in California to quantify fire impacts on surface fluxes and atmospheric composition and uses atmospheric models to understand how fires influence atmospheric chemistry, downwind ecosystems, and human health. Randerson received the James B. Macelwane Medal and Global Piers J. Sellers Mid-Career Award from the American Geophysical Union (AGU). He is a Fellow of the AGU and a member of the National Academy of Sciences. Randerson received a B.S. in chemistry and a Ph.D. in biological sciences from Stanford University. Randerson previously served on the National Academies of Sciences, Engineering, and Medicine's Committee on Methods for Estimating Greenhouse Gas Emissions and currently serves on the Committee on the Independent Study on Potential Environmental Effects of Nuclear War.

Randerson leads the science team for the Environmental Defense Fund's FireSat Exploratory Project.

Brendan Rogers is an associate scientist at the Woodwell Climate Research Center. He studies the vast expanses of boreal forests and Arctic tundra across Earth's northern high latitudes, with a particular interest in wildfires and permafrost ecosystems, including feedbacks to the global climate system. He combines field measurements, satellite remote sensing, and modeling to gain insight into rapidly changing carbon and energy cycles, vegetation dynamics, and disturbance regimes. Rogers uses his science to inform natural resource management and policies for improved climate mitigation, adaptation, and ecosystem protection and engages a range of stakeholders and rights holders, from local community members and fire managers to international policymakers, to explore the societal ramifications of his work. He is deputy lead for Permafrost Pathways, an initiative funded through the Audacious Project that addresses the local to global impacts of permafrost thaw. Rogers received an M.S. in environmental sciences from Oregon State University and a Ph.D. in Earth system science from the University of California, Irvine.

Rogers and Woodwell Climate Research Center have made public statements regarding wildfires, greenhouse gas emissions, and climate change.

Amber Soja is a physical scientist in the Chemistry and Dynamics Branch at the National Aeronautics and Space Administration (NASA) Langley Research Center, with a focus on Wildland Fire Science Program Management. She is currently serving as a manager for the NASA Applied Sciences Wildland Fire program. Her research uses Earth observations and models as tools to explore the dynamic interactive relationships between fire regimes, fire weather, air quality, the biosphere, atmosphere, and climate systems. Soja developed fire weather–specific fuel databases and fire emission inventories that are currently used in field campaigns and models. Her work includes integrating multiple satellite platforms in the Environmental Protection Agency's (EPA's) National Fire Emissions Inventory that supports EPA, state, and regional decisions. She has served on the National Science Foundation Wildfire & the Biosphere Innovation Lab panel, the Aerosol-Cloud Convection and Precipitation team for the selection of the Atmospheric Observing System mission, as a board member the International Association of Wildland Fire, and as a member on the Subcommittee on Disaster Reduction for the National Science and Technology Council development of "Wildland Fire Science and Technology Task Force Final Report." Soja received a B.A. and Ph.D. in environmental sciences from the University of Virginia.

Appendix C
Workshop Agenda

GREENHOUSE GAS EMISSIONS FROM WILDLAND FIRES: TOWARD IMPROVED MONITORING, MODELING, AND MANAGEMENT—A WORKSHOP
VIRTUAL | KECK CENTER (500 5th St. NW Washington DC, 20001)

WEDNESDAY, SEPTEMBER 13, 2023 | All times ET
Current Understanding of Biomes Vulnerable to Wildland Fires and Implications for GHG Emissions

10:30–10:40	Welcome and Opening Remarks **Loretta Mickley**, Harvard University, Committee Chair
10:40–10:50	Opening Remarks **Ann Bartuska** and **Steve Hamburg**, Environmental Defense Fund
10:50–11:45	Session 1: Framing the Workshop and Charge to Participants *Moderators: Don Hankins, CSU Chico, & Loretta Mickley, Harvard University, Planning Committee Members* *18-minute talks + 15-minute Q&A* **Nancy French**, Michigan Technological University, *Planning Committee Member* **Scott Stephens**, UC Berkeley
11:45–12:45	Session 2: Roles of Fire in Presently Vulnerable Biomes and the Associated Net GHG Emissions *Moderators: Nancy French, Michigan Technological University, & Loretta Mickley, Harvard University, Planning Committee Members* *15-minute talks + 15-minute Q&A* **Hélène Genet**, University of Alaska Fairbanks **Matt Hurteau**, University of New Mexico **Susan Page**, University of Leicester, and **Tom Smith**, London School of Economics
12:45–13:30	Lunch
13:30–14:20	Session 3: Management of Fires and Ecosystems and Implications for GHG Emissions: Recent Past and Current

Appendix C

     Moderators: Don Hankins, CSU Chico, & Nancy French, Michigan Technological University, Planning Committee Members
     5-minute lightning talks + 20-minute Q&A/discussion
     Amy Cardinal Christianson, Canadian Forest Service
     Dan Thompson, Canadian Forest Service
     Hugh Safford, UC Davis
     Karin Riley, U.S. Forest Service
     Bibiana Bilbao, Universidad Simón Bolívar
     Cynthia Fowler, Wofford College

14:20–14:30 Instructions for Breakout Discussions
14:30–14:45 Break

14:45–15:25 Session 4: Breakout Discussions
     Participants join assigned in-person or virtual breakout rooms
     Discussion moderators: **Sara Ohrel**, EPA; **Carly Phillips**, Union of Concerned Scientists; **Cynthia Whaley**, Environment and Climate Change Canada; **Ane Alencar**, Amazon Environmental Research Institute; **Joanne Hall**, University of Maryland; **Tatiana Loboda**, University of Maryland; **Adam Moreno**, CARB; **Douglas Morton**, NASA
     Discussion questions:
     - Which biomes and ecosystems merit the closest attention as climate changes and fire activity increases in some regions of the world? Has the workshop so far overlooked some vulnerable biomes/ecosystems?
     - Are there scalable and sustainable methods of land management that have not yet been discussed?
     - What uncertainties in fire activity or the carbon budget would be good to pin down? What are the major gaps in our knowledge of GHG emissions from fire that have not yet come to light?
     - What synergies and differences are there from biome to biome regarding GHG emissions? What lessons can we transfer from one region to another?

15:25–15:55 Report Back to Plenary
     5-minute transition back to plenary
     Breakout moderators report back on key takeaways from their discussions

15:55–16:00 Wrap Up and Plans for Day 2

16:00   END OF DAY 1

THURSDAY, SEPTEMBER 14, 2023 | All times ET
Observations and Modeling Needs and Opportunities

10:30–10:40	Welcome and Opening Remarks
Loretta Mickley, Harvard University, Committee Chair
10:40–10:50	Recap from Day 1
Loretta Mickley, Nancy French, Don Hankins, *Planning Committee Members*
10:50–12:40	Session 1: Observations of Wildland Fires and their GHG Emissions: Opportunities, Gaps, and Challenges
Moderators: Chris (Fern) Ferner, Johns Hopkins University, & Amber Soja, NASA, Planning Committee Members
10-minute talks + 40-minute Q&A/discussion
Andy Hudak, U.S. Forest Service
Jeff Vukovich, U.S. Environmental Protection Agency
Morgan Varner, Tall Timbers
Martin Wooster, King's College London
Bo Zheng, Tsinghua University
Louis Giglio, University of Maryland
12:40–13:25	Lunch
13:25–14:25	Session 2: Defining the Future: Modeling Challenges and Defining the Largest Data Gaps—Fire Emissions Inventories and Future Projections
Moderators: Amber Soja, NASA, & Jim Randerson, UC Irvine, Planning Committee Members
10-minute talks + 20-minute Q&A/discussion
Stijn Hantson, Universidad del Rosario, Bogotá
Park Williams, UCLA
Matthew Jones, University of East Anglia
Jed Kaplan, University of Calgary
14:25–14:35	Instructions for Breakout Discussions
14:35–14:50	Break
14:50–15:30	Session 3: Breakout Discussions
Participants join assigned in-person or virtual breakout rooms
Discussion moderators:

Appendix C 79

                Rebecca Scholten, Vrije Universiteit Amsterdam; **Carly Phillips**, Union of Concerned Scientists; **Cynthia Whaley**, Environment and Climate Change Canada; **Ane Alencar**, Amazon Environmental Research Institute; **Joanne Hall**, University of Maryland; **Thomas Buchholz**, Spatial Informatics Group; **Tina Liu**, UC Irvine; **Douglas Morton**, NASA
Discussion questions:
- Identify the largest gaps in data, observations, and/or models that can inform what is needed to reduce uncertainties in carbon emissions to increase the potential for successful mitigation outcomes?
- What are the data gaps and model enhancements that need to be improved to predict the consequences of various proposed mitigation actions on estimates of future fire GHG emissions (e.g., prescribed fire vs. wildfire over time)?
- What carbon stocks are most vulnerable to contemporary and future fire? Please consider the mass of potential loss and climate and socio-ecological impacts.

15:30–15:55   Report Back to Plenary
5-minute transition back to plenary
Breakout moderators report back on key takeaways from their discussions

15:55–16:00   Wrap Up and Plans for Day 3

16:00   END OF DAY 2

FRIDAY, SEPTEMBER 15, 2023 | All times ET
Future Management to Support Net-Zero Targets

09:00–09:10   Welcome and Opening Remarks
Loretta Mickley, Harvard University, Committee Chair

09:10–09:20   Recap from Day 2
Chris (Fern) Ferner, Jim Randerson, Amber Soja, *Planning Committee Members*

09:20–10:10   Session 1, Part 1: Wildfire Emission Impacts on National Reporting and Implications for Net-Zero Targets
Moderators: Sally Archibald, University of Witwatersrand, & Brendan Rogers, Woodwell Climate Research Center, Planning Committee Members

10-minute talks + 20-minute discussion
Werner Kurz, Canadian Forest Service
Grant Domke, US Forest Service
David Bowman, University of Tasmania

10:10–10:25 Break

10:25–11:40 Session 1, Part 2: Opportunities for Solutions to Reduce Future Wildfire Emissions in Different Biomes
10-minute talks + 25-minute discussion
Paul Hessburg, U.S. Forest Service
Peter Frumhoff, Harvard University/Woodwell Climate Research Center
Susan Page, University of Leicester
Marcia Macedo, Woodwell Climate Research Center
Geoff Cary, Australian National University

11:40–11:50 Instructions for Breakout Discussions

11:50–12:35 Lunch

12:35–13:20 Session 2: Breakout Discussions
Participants self-select in-person or virtual breakout rooms
Discussion moderators: **Rebecca Scholten**, Vrije Universiteit Amsterdam; **Sara Ohrel**, EPA; **Gyami Shrestha**, Lynker Corporation; **Merritt Turetsky**, University of Colorado Boulder; **Tatiana Loboda**, University of Maryland; **Elizabeth Wiggins**, NASA; **Marisol Maddox**, Wilson Center
Discussion questions:
- What are climate-effective, socially inclusive, and ecologically appropriate mitigation efforts to reduce future wildfire emissions?
- What are the barriers to implementation (trade-offs and co-benefits)?

13:20–13:35 Break

13:35–14:00 Report Back to Plenary
Breakout moderators report back on key takeaways from their discussions

14:00–15:15 Session 3: The Solution Space and Next Steps: Forest Management of Tomorrow and Livable Emissions

Moderators: Werner Kurz, Canadian Forest Service, & Brendan Rogers, Woodwell Climate Research Center, Planning Committee Members
Brief intros and moderated discussion
Paul Hessburg, U.S. Forest Service
Jimmy Fox, U.S. Fish and Wildlife Service
Dan Thompson, Canadian Forest Service
Natasha Ribeiro, University of Eduardo Mondlane College of Agriculture and Forestry, Maputo, Mozambique
Barry Hunter, Aboriginal Carbon Foundation
Jayaprakash Murulitharan, Cambridge University

15:15–15:25 Day 3 Synthesis
Sally Archibald, **Werner Kurz**, **Brendan Rogers**, *Planning Committee Members*

15:25–15:30 Closing Remarks
Loretta Mickley, Harvard University, Committee Chair

15:30 ADJOURN WORKSHOP